Library of Davidson College

ENERGY AND RESOURCE RECOVERY FROM WASTE

Energy and Resource Recovery from Waste

by

Stephen C. Schwarz, P.E.
Calvin R. Brunner, P.E.

Malcolm Pirnie, Inc.
White Plains, New York

NOYES DATA CORPORATION
Park Ridge, New Jersey, U.S.A.
1983

Copyright © 1983 by Stephen C. Schwarz and Calvin R. Brunner
No part of this book may be reproduced in any form
without permission in writing from the Publisher.
Library of Congress Catalog Card Number: 83-13110
ISBN: 0-8155-0959-6
ISSN: 0270-9155; 0090-516X
Printed in the United States

Published in the United States of America by
Noyes Data Corporation
Mill Road, Park Ridge, New Jersey 07656

10 9 8 7 6 5 4 3 2 1

Library of Congress Cataloging in Publication Data

Schwarz, Stephen C.
  Energy and resource recovery from waste.

  Bibliography: p.
  Includes index.
    1. Refuse as fuel. 2. Recycling (Waste, etc.)
  I. Brunner, Calvin R. II. Title.
  TP360.S38 1983   662'.8   83-13110
  ISBN 0-8155-0959-6

Energy Technology Review No. 86
Pollution Technology Review No. 102

To

My Wife Helena

　　—S.C. Schwarz

To

Bonnie E. Brunner
Jeryl S. Brunner
Renée E. Brunner

　　—C.R. Brunner

Preface

Natural resources within the United States do not exist in unlimited supply. As the population increases in size and in affluence, the demand for energy and raw materials increases at an alarming rate. No less a concern is the generation of waste and the need for its safe and economical disposal.

In this book, the relationship of these two trends are evaluated in a qualitative and quantitative manner. Current technology for the recovery of energy and other resources from solid waste is well proven. These technologies are presented and applied to various situations. Facilities or processes will not be implemented without an understanding of their economics and this book also presents cost data and financing alternatives.

Background material is presented which covers estimating solid waste generation, source separation techniques, evaluating the impact of a facility on the environment and a discussion of the legal aspects of resource recovery.

This book provides a complete and coordinated guide to waste generation and resource recovery techniques currently available, their selection, design and financing. It reflects current costs and up-to-date mechanisms for implementation of these facilities.

About the Authors

Stephen C. Schwarz, P.E., is Manager of Solid Waste Programs for Malcolm Pirnie, Inc., White Plains, New York. He is responsible for the planning and design of solid waste and resource recovery projects including landfills, refuse collection facilities and equipment, RDF systems and steam generating facilities. His activities include facilities permitting, financial consultation and the marketing of recoverable materials. Mr. Schwarz has a BS from CCNY, an MS from Manhattan College, is a Diplomate of the American Academy of Environmental Engineers, a member of the ASME Solid Waste Processing Division, ASCE, WPCF, and is licensed as a Professional Engineer in three states.

Calvin R. Brunner, P.E., is Chief Mechanical Engineer of Malcolm Pirnie, Inc. He has over 15 years experience in the design, construction and operation of resource recovery facilities for both government and industry and presents seminars on resource recovery and incineration throughout the United States. He has a BS from CCNY, MS from the Pennsylvania State University, is a registered Professional Engineer in five states and is a Diplomate of the American Academy of Environmental Engineers. He is an active member of the ASME Solid Water Processing Division and is a member of ASTM. Mr. Brunner has written the text *Design of Sewage Sludge Incineration Systems* (1980) published by Noyes Data Corp. He has also written technical articles for ASME, trade magazines and a series of articles for Chemical Engineering Magazine on computer programming.

Contents

1. INTRODUCTION ... 1
   Historical Perspective ... 1
   Energy Recovery Potential from Solid Waste ... 2
   Basis of Economic Feasibility ... 3

2. SOLID WASTE: QUANTITY AND COMPOSITION ... 6
   General ... 6
   Municipal Solid Waste ... 7
     Definition ... 7
     Estimating MSW Generation Rates ... 8
     Published Data ... 8
     Weight Data ... 8
     Volume Data ... 9
     Projecting MSW Quantities ... 10
   Industrial Wastes ... 11
     Definition ... 11
     Generation Rates ... 11
     Published Typical Data ... 11
     Weight Data ... 12
     Volume Data ... 12
     Projecting Industrial Solid Waste Quantities ... 13
   Other Wastes ... 14
   Seasonal Variations ... 14
   Solid Waste Composition ... 14
     General ... 14
     Typical Composition ... 14
     Moisture Content ... 15
     Heating Value ... 16
     Determination of Composition in the Field ... 17

Industrial Waste Composition ... 18
Statistical Analysis. ... 18
   Definitions. ... 18
   Methodology ... 22
References. ... 23

3. MARKETING RECOVERED ENERGY AND MATERIALS PRODUCTS ... 25
   General ... 25
   Potential Materials Markets. ... 25
      General ... 25
      Paper ... 26
      Ferrous Metals ... 27
         Copper Precipitation ... 27
         Remelt ... 27
         Detinning. ... 27
         Secondary Scrap Dealers. ... 27
      Aluminum ... 28
         Major Aluminum Producers. ... 28
         Secondary Aluminum Smelters ... 28
         Scrap Dealers ... 29
      Glass. ... 29
         Glass Container Manufacturers. ... 29
         Other Products ... 30
      Other Nonferrous Metals (Exclusive of Aluminum) ... 30
   Potential Energy Markets ... 31
      General. ... 31
      Solid Fuels. ... 32
         Fluff RDF ... 32
         Powder RDF ... 32
         Densified RDF ... 32
      Pyrolysis ... 32
      Methane Gas. ... 33
      Steam (or Hot Water). ... 34
      Electricity ... 34
   Product Specifications ... 35
      General. ... 35
      Paper ... 36
      Ferrous Metals ... 36
      Aluminum. ... 36
      Glass. ... 37
      Nonferrous Metals (Exclusive of Aluminum). ... 37
      RDF. ... 37
         Moisture Content ... 37
         Ash Content. ... 38
         Particle Size. ... 38
         Energy Yield ... 38
      Methane Gas. ... 38
      Steam ... 38

Estimating Revenues 39
   General. 39
   Ferrous Metals 39
   Other Non-Ferrous (ONF). 41
   Aluminum. 41
   Paper 41
   Glass. 42
   Typical Estimated Materials Products Revenues. 42
   Recovered Energy Products. 42
   Typical Estimated Energy Products Revenues 44
Market Securement 44
Sources of Information. 47
   General. 47
   Materials Purchasers. 47
   Energy Users 48
References. 48

4. **UNIT PROCESSES OF RESOURCE RECOVERY**. 50
   General 50
   Preprocessing 50
      Weighing Facilities. 50
      Receiving and Storage 53
         Pit and Crane 57
         Tipping Floor. 58
         Depressed Tipping Floor. 59
         Live Bottom. 59
      Trommel Screens. 63
      Shredding 68
      Air Classification. 80
   Energy Recovery Systems. 85
      Central Disposal Incineration Systems. 86
         European Systems. 86
         American Systems. 89
      Pyrolysis 91
      Controlled Air Incineration. 92
      Rotary Kiln 95
      Combustion Calculations 95
   Materials Recovery Unit Processes 99
      Handpicking. 100
      Screening. 100
         Trommels 100
         Vibrating Screens 101
      Shredders. 101
      Ferrous Material Recovery 101
      Glass Recovery 106
         Optical Sorting. 106
         Froth Flotation. 106
      Nonferrous Metal Recovery. 114

xiv   Contents

        Jigging...114
        Water Elutriation..114
        Heavy Media Separation..................................115
        Eddy Current Separation.................................115
        Electrostatic Separation..................................117
    Materials Handling Unit Process.............................117
        Apron Conveyors..117
        Belt Conveyors..121
        Pneumatic Conveyors....................................125
        RDF Storage Systems....................................126
            Atlas Silo..126
            Live Bottom Bins....................................126
            Crane and Bucket Retrieval...........................126
            Parascrew System....................................126
    References..130

5. RESOURCE RECOVERY SYSTEMS..........................131
    Introduction...131
    European vs. American Projects............................131
    Mass Burning...132
        General..132
        Process..132
        Status...133
    RDF Systems..133
        General..133
        Fluff RDF..135
        Densified RDF (d-RDF)..................................136
        Dust RDF..138
        Wet-Pulped RDF..140
    Pyrolysis Systems..140
        General..140
        Andco-Torrax..142
        Union Carbide Purox System.............................142
    Starved Air Systems......................................145
        General..145
    Status of Resource Recovery in the US......................150
    Recovery Efficiency......................................162
    Costs..162
    References..178

6. CODISPOSAL..179
    Introduction...179
    Codisposal Techniques...................................180
        General..180
        Incineration with Solid Waste.............................180
        Use of RDF in a Sludge Incinerator.......................181
            The Multiple Hearth Incinerator.......................181
            Fluidized Bed Incineration............................185

Contents   xv

	Current Status	187
	Economics	192
	References	193

7. SOURCE SEPARATION ... 194
Introduction ... 194
Collection Methods ... 195
    General ... 195
    Separate Truck ... 195
    Rack System ... 196
Costs ... 197
Public Education ... 201
References ... 201

8. ENVIRONMENTAL CONSIDERATIONS ... 202
General ... 202
Facility Siting ... 204
Traffic ... 205
Water Pollution ... 207
    General ... 207
    Process Wastewater ... 207
    Cooling Water ... 208
Air Pollution ... 210
    General ... 210
    Particulates ... 211
    Other Criteria Pollutants ... 212
    Hazardous Air Pollutants ... 213
    Control Technologies ... 214
        Baghouse Filters ... 215
        Scrubbers ... 216
        Electrostatic Precipitators ... 217
        Control Technology Selection ... 221
Aesthetics ... 222
    General ... 222
    Odor ... 222
    Noise ... 223
    Visual ... 224
Residue Disposal ... 224
Permits and Regulatory Requirements ... 226
    General ... 226
    NEPA ... 226
    Air Quality Regulations ... 227
        New Source Performance Standards ... 227
        Criteria Pollutants ... 229
        Non-Criteria Pollutants ... 230
        PSD Program ... 230
        The PSD Application ... 231
    Water Quality Regulations ... 232

xvi Contents

    Construction and Operating Permits 233
    References 234

9. INSTITUTIONAL FACTORS 235
Introduction 235
Solid Waste Flow Control 235
    General 235
    Contract 236
    Flow Control Laws 236
    Artificially Low Tipping Fees 237
Procurement Methodology 237
    General 237
    Procurement Alternatives 238
        Formal Bids 238
        Negotiated Procurement 239
Implementation Alternatives 240
    General 240
    Conventional 240
    Turnkey 241
    Full Service 241
    Full Service with Public Ownership 241
    Modified Full Service 241
Legal Constraints 242
    General 242
    Solid Waste Flow Control 242
    Competitive Bidding Requirements 242
    Long-Term Contracts 243
    Regulation of the Sale of Recovered Energy Products 243
PURPA 243
    Qualifying Facilities 244
Financing Alternatives 247
    General 247
    Project Financing Alternatives 247
    General Obligation Financing Alternatives 248
Funding Assistance 252
    General 252
    References 253

APPENDIX A: MARKET ASSESSMENT DOCUMENTS 254

APPENDIX B: RECOVERED MATERIALS SPECIFICATIONS 261

1
Introduction

HISTORICAL PERSPECTIVE

Resource recovery, defined as the recovery of energy or materials from waste, is not a new idea. As early as 1895 the City of Hamburg, West Germany was producing steam and electric power from the incineration of municipal refuse. In 1905, lights on the Williamsburg Bridge in New York City were powered by a refuse fired steam plant.

Until the early 1950's refuse incinerators had air cooled refractory lined combustion sections, and recovered energy using separate waste heat boilers. In 1954, the first modern water-tube wall incinerator was built in Berne, Switzerland. Today there are over 500 refuse fired energy systems in Europe. Resource recovery through incineration has progressed much more slowly in the United States with only about 30 systems currently in operation. A 1977 comparison of the percent of municipal solid waste converted to energy by country contrasts the status of this technology in the US and Europe:

Denmark	60%
Switzerland	40
Netherlands	30
Sweden	30
Germany	20
England	10
United States	1

The reasons for the slow growth of refuse to energy systems in the US are uncertain, but include lower energy costs, greater availability of land, and greater acceptance of

Energy and Resource Recovery from Waste

landfill as a disposal method. As energy costs rise, as landfills become more difficult and costly to site, and as more stringent regulations make landfill disposal more expensive, the popularity of resource recovery is certain to grow.

ENERGY RECOVERY POTENTIAL FROM SOLID WASTE

Energy recovery from solid waste cannot solve our energy problems, but it can contribute to the solution. Example 1-1 shows how a rough estimate of the potential contribution can be made. As shown, a reasonable estimate of the maximum practical contribution of energy recovery projects is 0.9 quads per year, or about 1.2 percent of total current US consumption. However, if this energy could be substituted for imported crude oil, currently running at about 13 quads per year, (one quad is 10^{15} BTU, one million billion BTU) it would lead to a reduction in imports of about 6.9 percent.

EXAMPLE 1-1

Calculate the potential contribution of resource recovery to the US's overall energy supply assuming 5 lbs per capita per day of MSW with an as received heating value of 4500 Btu/lb. Current US consumption is 78 quads/year of which 13 quads/yr are imported.

For the entire US population, 220 million people:

220×10^6 people $\times$ 5 lb/day $\times$ 365 day/yr $\times$ ton/2000 lb = 200×10^6 ton/year

Total 1985 MSW = 200 million tons

MSW Energy value = 4500 Btu/lb

Total potential recovery = $200 \times 10^6 \times 2000$ lb/ton $\times$ 4500 Btu/lb
 = 1.8×10^{15} Btu/yr
 = 1.8 quads/yr

If 50% of the population lives in urban areas where recovery is practical:

Total practical recovery = $1.8 \times 0.5 = 0.9$ quads/yr

Total U.S. consumption is 78 quads/yr; crude oil imports are 13 quads/yr.

$\frac{0.9}{78} = 1.2\%$ of total consumption

$\frac{0.9}{13} = 6.9\%$ of crude oil imports

Reducing our oil imports would also have a balance of payments impact. Assuming $40 per 40-gallon barrel as the price of imported oil, and 5.9 million Btu per barrel, a reduction of 0.9 quads in imports would lead to:

$$\frac{0.9 \times 10^{15}}{5.9 \times 10^6} = 153 \times 10^6 \text{ barrels/yr}$$

$$153 \times 10^6 \times \$40 = \$6.1 \text{ billion/yr}$$

BASIS OF ECONOMIC FEASIBILITY

Two fundamental concepts need to be understood if resource recovery is to be seen in the correct perspective.

- o The first and most important goal of any solid waste resource recovery project is solid waste disposal.
- o There is no "gold in garbage".

The first point is important in the design of resource recovery facilities. Although the recovery of energy and material are significant benefits, and are important to economic feasibility, it is still solid waste disposal that is the first purpose of a resource recovery facility. Solid waste will continue to be generated day after day, and must be disposed of day after day, regardless of whether the energy from the facility can be marketed, or indeed, whether the facility is even working at all. In this regard, a resource recovery facility is completely different from a power plant or factory where production can be adjusted to meet demand and market conditions. This concept also indicates the motivation for most resource recovery projects: someone (usually a municipality) must have a solid waste disposal problem.

The second concept is stated in reaction to the misleading attitude, expressed more than once in the mass media, that the resources in solid waste are a new form of "gold" to be sold to eager entrepreneurs. The reality is that solid waste is, by definition, garbage. If it had value it wouldn't be discarded.

Of course, the energy and materials in solid waste do have value, but the cost of extraction and recovery outweights this value. The economic feasibility of resource recovery should not, therefore, be based on making a profit from the sale of recovered products, but rather on offsetting a portion of the costs of disposal with these revenues, so that the net disposal cost for resource recovery is equal to or less than the alternative costs.

Recently, several factors have combined to enhance the economic feasibility of resource recovery:

- o More stringent landfill regulation.
- o Public opposition to new landfills.
- o Increases in the value of recovered energy products.

More stringent landfill regulation, due both to the Resource Conservation and Recovery Act of 1976 (RCRA), and to increased state regulatory activity, increases the costs of the landfill alternative by requiring more careful design and siting. Public opposition to new landfills, fanned by Love Canal and other such incidents, raises landfill costs by forcing siting in remote areas. In some cases, public opposition has made development of a new landfill impossible.

Increases in energy prices, from less than $3 per barrel for oil before 1973 to nearly $40 today, has had a major effect on the economics of resource recovery. In addition, one important reason to implement a resource recovery project is the desire to stabilize disposal costs in spite of increasing inflation. This is possible because energy costs have and are expected to continue to increase faster than prices in general. Thus, the increased revenues from recovered energy can offset increases in operating costs due to inflation, leading to stable net disposal costs. Example 1-2 demonstrates this effect.

EXAMPLE 1-2

Consider a 500 tpd resource recovery facility with the following annual costs/revenues:

Debt service	=	$5 million
O&M	=	3 million
Energy revenues	=	(4 million)
Materials revenues	=	(0.5 million)
Net annual cost		$3.5 million

1. Calculate breakeven tipping fee:

$$\frac{\$3{,}500{,}000/\text{year}}{500 \text{ ton/day} \times 365 \text{ day/year}} = \$19.18/\text{ton}$$

2. Project future tipping fees, if energy revenues increase at 15 percent, general inflation is 12 percent (12% increase in materials revenues) and the refuse load remains constant at 500 tpd.

	Annual Cost (revenue) in 10^6 $		
	Initial	*5 years	*10 years
Debt Service	5.0	5.0	5.0
O&M	3.0	5.3	9.3
Energy revenues	(4.0)	(8.0)	(16.2)
Materials revenues	(0.5)	(0.9)	(1.6)
Net annual cost	3.5	1.4	(3.5)
Tipping fee	$19.18/ton	$7.67/ton	($19.18/ton)

*An annual increase of 15% in cost C is equal to C x $(1.15)^n$ with n = number of years. For instance, energy revenues of $4.0 x 10^6 will increase, in 10 years, to $(4.0)(1.15)^{10}$ = $16.2 x 10^6..

2

Solid Waste: Quantity and Composition

GENERAL

Solid waste is the raw material of any resource recovery facility. The economics of resource recovery are sensitive to the quantity and composition of the solid waste received since these parameters affect the capital cost of the facility, as well as revenues from both tipping fees (user charge at the resource recovery facility) and the sale of recovered energy and materials products.

Example 2-1 illustrates the importance of correct estimates of solid waste quantities to facility economics. As shown, a 10 percent shortfall in solid waste (1000 to 900 tpd) can cause a 45 percent increase in tipping fees.

EXAMPLE 2-1

Consider a resource recovery facility designed for 1000 tpd. Assume the following:

Capital Cost =	$100 million
Debt Service =	$ 10 million/yr
Annual O&M =	$ 7 million
Energy Price (steam) =	$ 6/MMBTU
Materials Price (ferrous) =	$35/ton

1. Calculate Annual Energy Revenues assuming 4500 Btu/lb for MSW and 60% of this energy is recovered as steam.
   Energy input =
   1000 tpd x 365 x 2000 lbs/ton x 4500 Btu/lb = $3.29 \times 10^{12} \frac{Btu}{yr}$

Energy Recovered = 3.29×10^{12} Btu/yr $\times$ 0.60 recovery

$= 1.97 \times 10^{12}$ Btu/yr

Annual energy revenues = 1.97×10^{12} Btu/yr $\times$ \$$6/10^6$ = 11.8×10^6

2. Calculate Annual Materials Revenues assuming MSW contain 8% ferrous and 90% of this material can be recovered:

Ferrous input = 1000 tpd $\times$ 365 $\times$ 0.08 ferrous = 29,200 tpy

Ferrous Recovered = 29,200 $\times$ 0.9 recovery = 26,280 tpy

Annual materials revenues = 26,280 tpd $\times$ \$35/ton = \$918,000

3. Calculate Tipping Fee
   - Debt Service = $\$10 \times 10^6$/yr
   - O & M = 7×10^6
   - Energy = (11.8×10^6)
   - Materials = (0.9×10^6)
   - Net Annual Cost = $\$4.3 \times 10^6$
   - Tipping Fee = $\dfrac{\$4.3 \times 10^6}{1000 \text{ tpd} \times 365}$ = \$11.78/ton

4. Assume 10% Shortfall in solid waste. Recalculate Tipping Fee:
   - Debt Service = $\$10 \times 10^6$/yr (constant)
   - O & M = 7×10^6 (constant)
   - Energy $0.9 \times 11.8 \times 10^6 = (10.6 \times 10^6)$
   - Materials $0.9 \times 0.9 \times 10^6 = (0.8 \times 10^6)$
   - Net Annual Cost = $\$5.6 \times 10^6$
   - Tipping Fee = $\$5.6 \times 10^6$/900 tpd $\times$ 365 = \$17.05/ton

This is an increase of $\dfrac{17.05 - 11.78}{11.78} \times 100$ = 45% in the tipping fee at the design throughput.

Another point this example illustrates is the quasi-business nature of a resource recovery facility, as opposed to the service nature of a conventional tax-financed, public works project such as a highway, sewage treatment plant or school. Most resource recovery facilities must pay their own way by charging user fees and by sale of recovered energy and materials: a facility which cannot charge a reasonable fee is not feasible.

MUNICIPAL SOLID WASTE

Definition

Municipal solid waste (MSW) is generally defined as the combined residential, commercial, and institutional, solid wastes generated in a given municipal area.[1] It includes all wastes normally collected from residences, small businesses (retail stores, restaurants, markets, office buildings, hotels, print shops, auto repair shops, and the like) and institutions (schools, hospitals, and the like). Not included are wastes generated by sizable manufacturing enterprises (see Industrial Wastes, below).

Estimating MSW Generation Rates

MSW is generally the predominant waste type handled at a resource recovery facility, and an accurate estimate of quantity is correspondingly important. Estimates are generally made based on generation rates (lbs per capita per day or lb/cap/day) and population data. Generation rates can be obtained in three ways:
- Published data
- Weight data records
- Volume data records

Published Data

Studies of MSW generation have been performed in many communities and typical data on generation rates collected. Results have ranged from 2.0 to 5.0 lb/cap/day.

Probably the best discussion of this issue is given by Frank A. Smith, an USEPA economist,[2] who concludes that the best estimate for the residential fraction of MSW lies between 2.3 to 2.7 lbs/cap/day; for the commercial-institutional fraction he estimates 1.6 to 1.7 lbs/cap/day. These figures lead to a total MSW generation rate of 3.9 to 4.4 lbs/cap/day.

Based on other studies, and on the need to be conservative in evaluating resource recovery feasibility, it is suggested that 3.5 lbs/cap/day be used for total residential, commercial and institutional generation, in the absence of more specific local data.

Use of typical numbers for estimating solid waste generation is advisable only where other data is not available, and then only for preliminary feasibility evaluation. Solid waste generation varies too much from community to community to permit the use of such generalized data for facility sizing and design.

Weight Data

The most accurate method of determining solid waste

quantities is to weigh the collection vehicles at a convenient point such as an incinerator, landfill, or transfer station. This can be accomplished either with permanent platform type truck scales if the facility is so equipped, or with temporary, rented wheel or axle scales. Example 2-2 shows how solid waste generation rates may be estimated from such data.

It should be noted that every truck serving a particular municipality for a given period of time must be weighed.

EXAMPLE 2-2

Vehicle Serial No.	Vehicle Type	Weight
001	Compactor	10,500
002	Compactor	9,000
003	Compactor	9,200
004	Roll-off	6,100
005	Pickup	100
		34,900 lbs/day

If population is 10,000, then per capita generation = 3.49 lb/cap/day

Volume Data

Many municipal and private solid waste disposal facilities are not equipped with scales and charge for disposal by volume. If such volume data is available it can be used to estimate solid waste quantities using estimated or measured solid waste densities.

In the absence of other data, the density of MSW in a compactor-type collector truck can be assumed to be 500 lbs/cu yd; in a non-compacting type truck 150 lbs/cu yd; in an individual private vehicle 100 lbs/cu yd.

These densities all assume each collection vehicle is full. Since this is often not the case it is advisable to confirm these densities with a limited on-site weighing program. Example 2-3 illustrates this method.

EXAMPLE 2-3

1. Weigh sample of vehicles, assuming that each vehicle is full.

Vehicle	Volume	Net Weight*	Density
Compactor	20 cu yds	10,500 lbs	525 lbs/ cu yd
Compactor	18	9,000	500
Compactor	20	9,200	460
Roll off	40	6,100	153
Pickup	1	100	100

2. Calculate Typical Densities:

$$\text{Compactor} = \frac{(10{,}500 + 9{,}000 + 9{,}200) \text{ lb}}{(20 + 18 + 20) \text{ cu yd}} = 495 \text{ lbs/cu yd}$$

$$\text{Roll off} = \frac{6{,}100 \text{ lb}}{40 \text{ cu yd}} = 153 \text{ lbs/cu yd}$$

$$\text{Pickup} = \frac{100 \text{ lbs}}{1 \text{ cu yd}} = 100 \text{ lbs/cu yd}$$

3. Use typical densities and volume data to calculate solid waste generation:

Vehicle	Volume	Loads Daily	Density	Weight
Compactor	20 cu yds	10	495	99,000 lbs
Compactor	18	20	495	178,200
Rolloff	40	31	153	189,720
Rolloff	43	26	153	171,054
Pickup	1	6	100	600
				638,574

4. Calculate Per Capita Generation if population is 200,000:

638,574/200,000 = 3.19 lbs/cap/day

*Net weight is the weight of the loaded truck less its tare weight where tare is the unloaded truck weight.

Projecting MSW Quantities

MSW quantities may be projected by multiplying per capita generation rates by projected population estimates which are generally available from local planning agencies. Per capita solid waste generation is not constant, and in recent years has tended to increase steadily by up to 2 percent annually in the US.

Recent economic trends have led to a leveling off of generation of solid waste. Although insufficient data exists to calculate or project the rate of increase in per capita generation, it is believed that an allowance of 1 percent per year should be made for this growth in the future.

Example 2-4 illustrates the projection of MSW quantities.

EXAMPLE 2-4

	1980		1990		
Town	Per Capita MSW generation	Population	Per Capita MSW Generation	Estimated Population	TPD
Bedford	3.19	200,000	3.52	220,000	387
North Salem	3.05	51,000	3.37	59,000	99
Scarsdale	4.10	34,000	4.53	65,000	147
North Castle	4.51	158,000	4.98	180,000	448
Harrison	2.93	75,000	3.24	78,000	126
					1207

Sample Calculation:
For 10% increase in generation rate per year for 10 years:

Bedford: $3.19 \times (1.01)^{10} = 3.52$ lb per capita per day

$3.52 \times \dfrac{220,000}{2000} = 387$ tons/day

INDUSTRIAL WASTES

Definition

Industrial wastes are those waste materials discarded from industrial operations or derived from manufacturing operations, including construction, fabrication, light and heavy manufacturing, refineries, chemical plants, lumbering, mining, power plants, and so forth.

Generation Rates

The 1968 National Survey of Community Solid Waste Practices[3] developed a national average industrial solid waste generation rate of 1.86 lbs/cap/day. Other sources have suggested a range of 1.0 to 3.5 lbs/cap/day. In fact, industrial waste composition and quantity vary so much from locality to locality that such generalized numbers are not realistic except for estimating national quantities. There are no reliable means to determine industrial solid waste quantities except to examine each facility separately.

Published Typical Data

For industrial wastes, it is more useful to present generation rates in terms of 1000 lbs per employee per year (Kpey) than in lbs per capita. Table 2-1 presents results from two surveys of typical industrial waste generation rates utilizing the standard Industrial Code (SIC) classification system. The FMC study covered only California, while the Combustion Engineering study was nationwide. This table represents a guide to industrial solid waste production and can be used with local data on numbers of employees and industry types to formulate a preliminary estimate of industrial waste quantities.

TABLE 2-1

INDUSTRIAL SOLID WASTE PRODUCTION FACTORS (1965)

Standard Industrial Classification Code		FMC Study (Kpey)	Combustion Engineering Study (Kpey)
No.	Title		
19.	Ordinance & Accessories	1.3	2.9
20.	Food Processing	10.4	10.8
21.	Tobacco	-	10.8
22.	Textile	-	2.5
23.	Apparel	1.0	0.6
24.	Lumber & Wood Products	43.4	160 (total) 44 (less sawmills)
25.	Furniture & Fixtures	40.3	10.2
26.	Paper & Allied Products	25.1	17.5
27.	Printing, Publishing, & Allied	26.4	16.5
28.	Chemicals & Allied	16.4	12.6
29.	Petroleum Refining	-	-
30.	Rubber & Plastics	3.1	11.9
31.	Leather	-	19.4
32.	Stone, Clay, Glass, & Concrete	36.2	5.0 (glass)
33.	Primary metals	-	3.0
34.	Fabricated Metal Products	13.5	2.9
35.	Non-electrical Machinery	8.4	2.9
36.	Electrical Machinery	6.0	2.1
37.	Transportation Equipment	6.8	2.1
38.	Instruments	5.0	5.4
39.	Miscellaneous Manufacturing	5.0	4.6
All manufacturing		-	10.5
All manufacturing (less sawmills)			6.5

Source: Reference 4

Weight Data

Most industries must pay for solid waste disposal and therefore keep careful records of quantities and costs. Where this data is available on a weight basis it should be used as the best available estimate of waste quantities.

Volume Data

If only volume data on industrial waste quantities are

available, this data must be converted to weights following a methodology similar to that used for MSW. Table 2-2 presents typical density data for various industrial solid wastes.

TABLE 2-2
INDUSTRIAL SOLID WASTE
DENSITY DATA

Waste	lbs/cu yd As Discarded
Dept. Store Waste	80
Hospital Waste (not research)	100
School Waste w/lunch program	110
Supermarket Waste	100
Bakelite	600
Bitumen Waste	1,500
Brown Paper	135
Cardboard	180
Cork	320
Corn Cobs	300
Corrugated Paper (loose)	100
Disposable Hospital Plastics	120
Grass, Green	75
Hardboard	900
Latex	1,200
Magazines	945
Meat Scraps	400
Milk Cartons, Coated	80
Nylon	200
Paraffin - Wax	1,400
Plastic Coated Paper	135
Polyethylene Film	20
Polypropylene	100
Polystyrene	175
Polyurethane (foamed)	55
Resin Bonded Fiberglass	990
Rubber - Synthetics	1,200
Shoe Leather	540
Tar Paper	450
Textile Waste (non synthetic)	280
Textile Waste (synthetic)	240
Vegetable Food Waste	375
Wax Paper	150
Wood	300

The above chart shows the various weights of materials commonly encountered in incinerator applications. The values given are approximate and may vary based on their exact characteristics or moisture content.

Projecting Industrial Solid Waste Quantities

Projection of industrial solid waste quantities are made in the same way as MSW projections: current generation rates per employee are estimated from available weight or volume data, and are then applied to estimated future numbers of employees. Data on growth in employment must be obtained from local industries or planning agencies.

OTHER WASTES

Municipal and industrial wastes comprise the bulk of material likely to be processed at a resource recovery facility. Other waste categories exist and should be considered in overall solid waste planning. Data on such wastes is summarized in Table 2-3.

TABLE 2-3
SELECTED PER CAPITA SOLID WASTE GENERATION RATES

Construction Debris	0.66 lb/cap/day
Street Sweepings	0.25
Tree and landscape	0.18
Park and beach	0.16
Catch basin	0.04
Sewage treatment plant solids	0.20
Scrap automobiles	0.25

SEASONAL VARIATIONS

Quantities of solid waste vary with the season of the year, due to variation in levels of economic activity, quantities of lawn clippings and leaves, and seasonal population changes.

For municipal solid waste, a variation of ±25 percent should be assumed, in the absence of long-term local data. For industrial wastes, data on local industrial employment patterns must be used to estimate seasonal variations.

SOLID WASTE COMPOSITION
General

The revenues earned by a resource recovery facility depend on the quantity of energy and materials recovered, which, in turn depend on the efficiency of the recovery process, and the amount of energy and recoverable materials in the input solid waste. For these reasons an estimate of solid waste composition is needed in the process of resource recovery implementation.

Typical Composition

Resource recovery economics are not as sensitive to variations in solid waste composition as to variations in quantity. For this reason, use of published data will usually suffice for feasibility determinations and preliminary design. Actual testing of waste to determine its composition is advisable before final design, and as a part of determining facility performance. Table 2-4, derived from Reference 5, presents a summary of the typical physical composition of MSW.

TABLE 2-4
TYPICAL PHYSICAL COMPOSITION OF MSW

	Percent by weight	
Component	Range	Typical
Food wastes	6-26	15
Paper	25-45	40
Cardboard	3-15	4
Plastics	2-8	3
Textiles	0-4	2
Rubber	0-2	0.5
Leather	0-2	0.5
Garden trimmings	0-20	12
Wood	1-4	2
Glass	4-16	8
Nonferrous metals	0-1	1
Ferrous metals	2-10	8
Dirt, ashes, brick, etc.	0-10	4

Moisture Content

The moisture content of solid waste is an important parameter in determining both heating value and density. Typical moisture contents are summarized in Table 2-5.

TABLE 2-5
TYPICAL MOISTURE CONTENT OF
MUNICIPAL SOLID WASTE COMPONENTS

	Moisture, percent	
Component	Range	Typical
Food wastes	50-80	70
Paper	4-10	6
Cardboard	4-8	5
Plastics	1-4	2
Textiles	6-15	10
Rubber	1-4	2
Leather	8-12	10
Garden trimmings	30-80	60
Wood	15-40	20
Glass	1-4	2
Tin cans	2-4	3
Nonferrous metals	2-4	2
Ferrous metals	2-6	3
Dirt, ashes, brick, etc.	6-12	8
Municipal solid wastes	15-40	20

Source: Reference 5

Heating Value

The energy content or heating value of solid waste is of critical importance in estimating the potential for energy recovery. Typical data on the heating value of MSW components is summarized in Table 2-6, adapted from Reference 5.

TABLE 2-6

HEATING VALUE OF MSW COMPONENTS

Component	Energy, Btu/lb	
	Range	Typical
Food wastes	1,500- 3,000	2,000
Paper	5,000- 8,000	7,200
Cardboard	6,000- 7,500	7,000
Plastics	12,000-16,000	14,000
Textiles	6,500- 8,000	7,500
Rubber	9,000-12,000	10,000
Leather	6,500- 8,500	7,500
Garden trimmings	1,000- 8,000	2,800
Wood	7,500- 8,500	8,000
Glass	50- 100	60
Tin cans	100- 500	300
Nonferrous metals	---	---
Ferrous metals	100- 500	300
Dirt, ashes, brick, etc.	1,000- 5,000	3,000
Municipal solid wastes	4,000- 5,500	4,500

The values in Table 2-6 are on an as-discarded basis. They may be converted to a dry basis by using the equation below:

$$\text{Btu/lb (dry basis)} = \text{Btu/lb (as-discarded)} \left(\frac{100}{100 - \% \text{ moisture}}\right)$$

Similarly, the energy content values may be converted to an ash-free dry basis:

$$\text{Btu/lb (ash-free dry)} = \text{Btu/lb (as-discarded)} \left(\frac{100}{100 - \% \text{ ash} - \% \text{ moisture}}\right)$$

Example 2-5 demonstrates how the energy content of a municipal solid waste sample may be estimated using the data in the above tables.

EXAMPLE 2-5

Given the typical physical composition of MSW stated below and the energy content in Table 2-6, estimate the total energy value:

Component	Solid wastes, lb	Energy Btu/lb	Total energy, Btu
Food Wastes	15	2,000	30,000
Paper	40	7,200	288,000
Cardboard	4	7,000	28,000
Plastics	3	14,000	42,000
Textiles	2	7,500	15,000
Rubber	0.5	10,000	5,000
Leather	0.5	7,500	3,750
Garden trimmings	12	2,800	33,600
Wood	2	8,000	16,000
Glass	8	60	480
Tin cans	6	300	1,800
Nonferrous metals	1	---	---
Ferrous metals	2	300	600
Dirt, ashes, brick, etc.	4	3,000	12,000
Total	100		476,230

$$\text{Energy content} = \frac{476{,}230 \text{ Btu}}{100 \text{ lb}} = \frac{4{,}762 \text{ Btu}}{\text{lb}}$$

Determination of Composition in the Field

Sampling and testing solid waste is difficult because of its heterogeneous nature. Most commonly used methods involve dumping a truckload of material in an enclosed area, and then reducing the truckload to an approximately 200 lb sample by a process of repeatedly quartering the waste, and discarding three fourths. Generally landfill equipment (a bulldozer or wheeled loader) is used for quartering. Alternative methods include selecting a random sample using a grid method, and treating the entire truckload as the sample.

After sample selection is completed, the refuse is then sorted into standard components (see Table 2-4) and each component weighed.

To determine heating value a representative 1 to 2 lb sample of the combustible fraction is reconstituted, and milled to uniform consistency. One gram of this sample is then tested in a bomb calorimeter.

The most accurate means of determining the heating value of MSW is to burn it in a waste fired boiler (an incinerator with heat recovery) and measuring the steam flow generated from a measured weight of MSW over a period if time, i.e., at least four hours. The steam generation when burning MSW is compared to the steam produced when burning supplemental fuel (natural gas or fuel oil). With appropriate corrections made for radiation and other losses the heating value of the waste can be readily calculated from the heat represented by the steam produced.

Industrial Waste Composition

Industrial wastes are more variable than MSW, and actual data on industry type and waste composition are therefore more important.

Certain data on industrial waste components and energy contents are listed in Tables 2-7, 2-8 and 2-9.

STATISTICAL ANALYSIS
Definitions

Many of the important variables in solid waste generation and composition are impossible to know or measure exactly. It is often useful to use simple statistical techniques to manage such data.

Certain commonly used statistical measures must be defined:

Frequency is the number of times a given value occurs in a set of observations.

Mean, also the arithmetic mean or average, is given by:

$$\overline{X} = \frac{\Sigma X}{N}$$

Where
$\overline{X}$ = mean
ΣX = summation of X values
N = number of observations

Median is the value such that half of all observations are below and half above.

Standard deviation is a useful measure of dispersion, given by:

$$s = \sqrt{\frac{\Sigma(\overline{X}-X)^2}{N-1}}$$

Where $\overline{X}$ = mean
N = number of observations

Coefficient of Variation is a measure of the relative variation of dispersion, given by:

$$CV = \frac{100\,\sigma}{\overline{X}}$$

Where CV = coefficient of variation, percent
σ = standard deviation
$\overline{X}$ = mean

TABLE 2-7
CLASSIFICATION OF WASTES TO BE INCINERATED[6]

Classification of Wastes Type	Description	Principal Components	Approximate Composition (% By Weight)	Moisture Content (%)	Incombustible Solids (%)	BTU Value/lb. of Refuse as Fired
Class -	Plastic	100% combustible plastic-all types	100% plastic	0	0	15,000
Class 0	Trash	Highly combustible waste. Paper, wood, cardboard cartons and up to 10% treated papers, plastic or rubber scraps: commercial and industrial sources	Trash 100%	10%	5%	8500
Class 1	Rubbish	Combustible waste, paper, cartons, rags, wood scraps, combustible floor sweepings: domestic, commercial and industrial sources.	Rubbish 80% Garbage 20%	25%	10%	6500
Class 2	Refuse	Rubbish and garbage: residential sources	Rubbish 50% Garbage 50%	50%	7%	4800
Class 3	Garbage	Animal and vegetable wastes: restaurants, hotels, markets, institutional, commerical, and club sources.	Garbage 65% Rubbish 35%	70%	5%	2500
Class 4	Animal solid and organics	Carcasses, organs, solid organic wastes: hospital, laboratory, abattoirs, animal pounds and similar sources	Animal and human tissue 100%	85%	5%	1000
Class 5	Gaseous, liquid or semi-liquid	Industrial process wastes incinerate directly through a burner.	Variable	Dependent on major components	Variable	Variable
Class 6	Semi-solid and solid	Combustibles requiring rotary retort equipment.	Variable	Dependent on major components	Variable	Variable

TABLE 2-8

INDUSTRIAL WASTE COMPOSITION BY SIC CLASSIFICATION[7]

SIC Code		Component (% By Weight)								
		Paper	Wood	Leather	Rubber	Plastics	Metals	Glass	Textiles	Food	Misc
20	Food & Kindred Products	52.3	7.7	–	–	0.9	8.2	4.9	0	16.7	9.2
22	Textile Mills	45.4	–	0	–	4.7	0	0	26.8	–	–
23	Apparel Products	55.9	–	0	0	–	–	0	36.5	1.35	–
24	Wood Products	16.7	71.6	0	0	0	–	0	0	0	7.8
25	Furniture	24.7	42.1	0	–	–	–	–	–	–	–
26	Paper & Allied Products	56.3	11.3	0	0	–	9.4	0	–	–	14.0
27	Printing & Publishing	84.9	5.5	–	0	–	–	0	–	–	–
28	Chemical & Allied	55.0	4.5	–	–	9.3	7.2	2.2	–	–	19.7
29	Petroleum & Allied	72.1	6.8	0	0	15.3	4.4	0	0	–	1.0
30	Rubber & Plastics	56.3	5.2	0	9.2	13.5	–	0	–	–	–
31	Leather	6.0	3.9	53.3	–	–	13.5	–	0	0	–
32	Stone, Clay & Glass	33.8	4.3	0	–	–	8.1	12.8	–	0	40.0
33	Primary Metals	41.0	11.6	0	–	5.4	5.5	2.0	0	–	29.0
34	Fabricated Metals	44.6	10.3	0	–	–	23.2	–	–	–	12.2
35	Nonelectrical Machinery	43.1	11.4	–	–	2.5	23.7	–	0	–	–
36	Electrical Machinery	73.3	8.3	0	–	3.5	2.3	–	0	1.2	19.5
37	Transporation	50.9	9.4	0	1.4	2.1	–	–	0	–	–
38	Scientific Instruments	44.8	2.3	0	0	6.0	8.4	–	0	–	–
39	Misc. Manufacturing	54.6	13.0	–	–	11.9	5.0	–	–	–	8.1

TABLE 2-9

ENERGY CONTENT OF COMMON INDUSTRIAL WASTES

Waste	Heating Value (Btu/lb-as discarded)
Acetic Acid	6,280
Animal Fats	17,000
Bakelite	12,500
Benzene	18,210
Bitumen Waste	16,570
Brown Paper	7,250
Carbon	14,093
Cardboard	6,810
Cork	11,340
Corn Cobs	8,000
Cotton Seed Hulls	8,600
Coffee Grounds	10,000
Citrus Rinds	1,700
Coated Milk Cartons	11,330
Corrugated Paper (loose)	7,040
Disposal Hospital Plastics	12,200
Ethyl Alcohol	13,325
Hydrogen	61,000
Hardboard	8,170
Kerosene	18,900
Latex	10,000
Linoleum Scrap	11,000
Magazines	5,250
Methyl alcohol	10,250
Meat Scraps	7,623
Naphtha	15,000
Nylon	13,620
Plastic Coated Paper	7,340
Polyethylene	20,000
Polyurethane (foamed)	13,000
Polystyrene	17,700
Rags (silk or wool)	8,400-8,900
Rags (linen or cotton)	7,200
Resin bonded fibre glass	19,500
Rubber - Sythetics	14,610
Shoe leather	7,240
Tar or asphalt	17,000
Tar paper	11,500
Turpentine	17,000
Textile Waste (non-synthetic)	8,000
Textile Waste (Synthetic)	15,000
Wax paraffin	18,621
Wax paper	11,500
Wood (average)	9,000
Wood sawdust (pine)	9,600
Wood bark (fir)	9,500

The coefficient of variation in solid waste generation rates is typically 10 to 60 percent, compared to 10 to 30 percent in biological treatment and 2 to 10 percent in chemical analyses. This indicates considerable dispersion, or variability, in solid waste data.

Methodology

The parameters defined above can be determined analytically, or graphically using special graph paper called arithmetic probabililty paper. Example 2-6 shows how this method may be used to analyze solid waste generation data.

EXAMPLE 2-6

The following data represents daily solid waste collection. Determine the capacity of a transfer station to handle this solid waste, if the capacity of the station is not to be exceeded more than 5 percent of the time.

Observation	Collection (tpd)
1	338
2	356
3	369
4	423
5	508
6	494
7	412
8	421
9	435
10	397
11	373
12	336

1. Arrange the observations in ascending order (rank) and calculate the plotting position [PP=(m/N+1)x100] for each observation. For this example N=12, the number of observations.

Rank Serial Number, m	Collection tpd	Plotting Position %
1	336	7.7
2	338	15.4
3	356	23.1
4	369	30.8
5	373	38.5
6	397	46.2
7	412	53.8
8	421	61.5
9	423	69.2
10	435	76.9
11	494	84.6
12	508	92.3
	$\Sigma = \overline{4862}$	

2. Calculate the mean:

$$\text{Mean} = \frac{\Sigma X}{N} = \frac{4862}{12} = 405.2 \text{ tpd}$$

3. Plot data on arithmetic probability paper (see Figure 2-1). Plot mean at 50 percent probability.

4. Plot a visual best fit line through the mean. The degree to which the data fits the line is a measure of the normality of the distribution.

5. For a normal distribution 68.27 percent of all observations fall within ± one standard deviation from the mean. The standard deviation can therefore be determined from the plot by subtracting the mean from the value at 84.1 percent (50% + 68.27/2):

   Value at 84.1% = 464.0 tpd
   Mean   405.2

   Standard deviation = 58.8 tpd

6. The required capacity is read from the plot at 95 percent

   Capacity = 518 tpd (128 percent of the mean)

REFERENCES

1. National Center for Resource Recovery, Inc. *Glossary of Solid Waste Management And Resource Recovery*

2. F.A. Smith, *Comparative Estimates of Post-Consumer Solid Waste*, EPA/530/SW-148 May 1975.

3. Muhich, A.J., A.J. Klee, and P.W. Britton, USDHEW, *Preliminary Data Analysis, 1968 National Survey of Community Solid Waste Practices*, ASCE Journal of the Sanitary Engineering Div. Vol. 96, No 5A, 1970.

4. Ross E. McKinney, Unpublished text.

5. Tchobanoglous, G., Theisen, H, and Eliassen, R; *Solid Wastes: Engineering Principals and Management Issues*; McGraw-Hill, 1977.

6. Incinerator Institute of America, *Incinerator Standards*, March 1970.

7. Wilson, D.G., *Handbook of Solid Waste Management*, 1977.

24   Energy and Resource Recovery from Waste

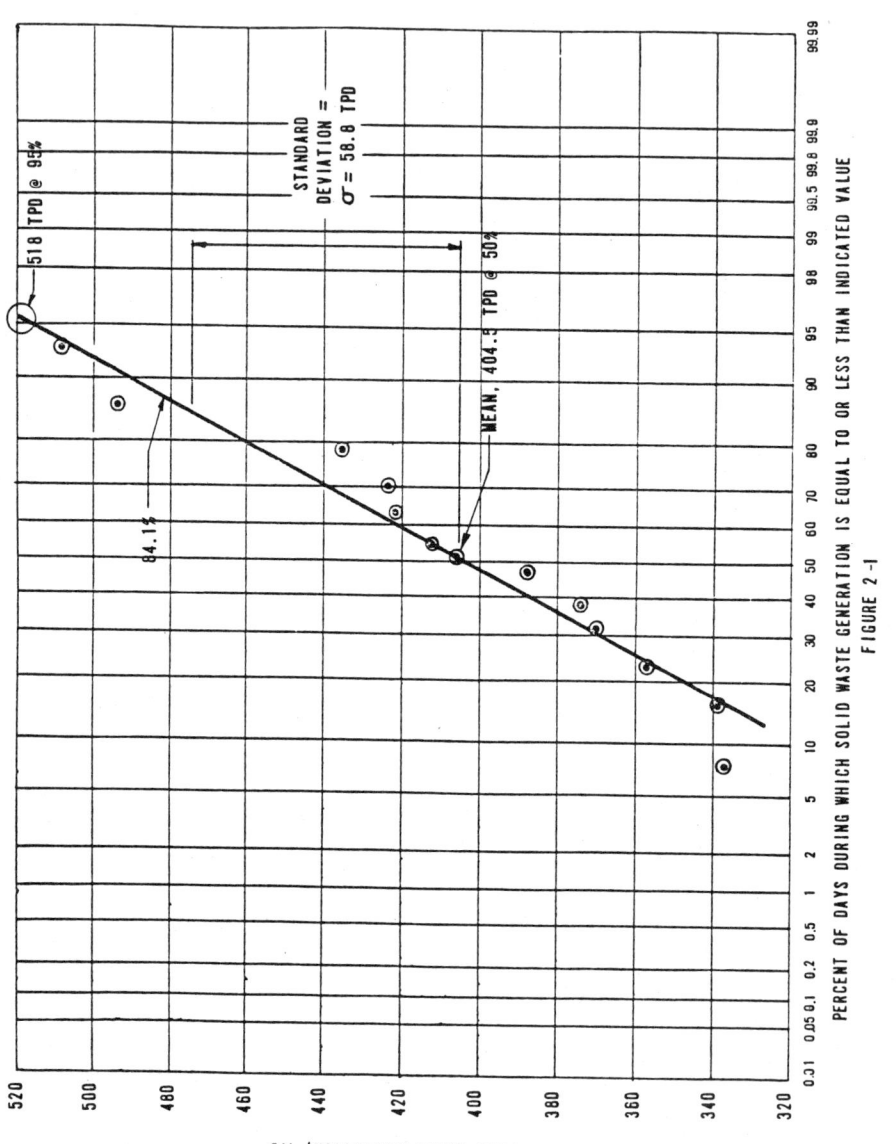

FIGURE 2-1

3
Marketing Recovered Energy and Materials Products

GENERAL

This chapter focuses on marketing the materials and energy products recovered from solid waste. Marketing should be an initial step in the resource recovery planning and implementation process since the markets for materials and energy will dictate the type of recovery technology selected. A comprehensive market assessment will include:

- o Identification of potential markets and local purchasers for materials and energy;
- o Product specifications;
- o Estimating revenues; and
- o Market securement

Each of these market assessment components are discussed below. In addition, to assist in identifying and locating potential purchasers a discussion of information sources on potential markets is included.

POTENTIAL MATERIALS MARKETS
General

Materials in municipal solid waste that can be recovered include: paper, ferrous metals (light and heavy fractions), aluminum, other non-ferrous metals and glass. The discussion of these materials will center on a definition of the materials (where appropriate), a brief description of recovery methods and a review of potential markets and specific purchasers.

Paper

Paper is generally the largest component of a municipal solid waste stream: it is composed of newsprint, corrugated boxes, and other mixed paper.

Newsprint and corrugated boxes can be recovered through source separation with some recovery possible via hand sorting of mixed waste at resource recovery plants. Mechanical fiber recovery processes have been developed; however, the performance of these systems in full sized resource recovery facilities is presently unproven. One system that has been employed in a full scale facility is the Fiberclaim system developed by Black Clawson, Inc. This system is used at the Hempstead, New York resource recovery facility to produce fuel for a waterwall combustion unit. In this sense it provides energy recovery and not materials recovery, since the paper fiber is burned and not reused in a paper product. This is similar to the manner in which other mixed paper is handled, since it is not generally recovered but is left in the waste stream for its fuel value.

The two largest markets for recovered paper are the paperboard and paper segments of the paper industry. Paperboard includes corrugated boxes, boxboard, linerboard and chipboard, while paper includes newsprint, printing paper and Kraft packaging. An uncontaminated source separated paper product is preferred since both markets are sensitive to contamination and moisture content. The degradation of fibers by moisture is a major problem in the use of recycled paper. To minimize this problem, paper which is to be reused should be bundled and separated from the waste stream to avoid contamination. These bundles can then be removed prior to processing of the waste.

Specific purchasers fall into two major categories: paper mills, which utilize the waste paper to produce a spectrum of paper products and local paper recyclers or scrap dealers, who serve as an intermediate step between the paper mills and those generating the waste paper. A listing of the paper mills located in the U.S. as of 1976 is provided in the USEPA publication <u>Market Locations for Recovered Materials</u>. Local scrap dealers may be found in the local telephone directory (Yellow Pages).

Ferrous Metals

Two marketable ferrous metal products can be extracted from MSW: a light ferrous fraction (mainly "tin" cans) and miscellaneous heavy ferrous scrap (hardware, rods, bars, pipe, castings, etc.). Cans are the predominent form of ferrous metals extracted from MSW.

Briefly, the technology for recovering ferrous metals is based on magnetic attraction. It has been demonstrated as an operable technology in numerous facilities and is a standard component in most resource recovery systems. For a detailed discussion of the ferrous recovery systems available, see Chapter 4.

There are four principal markets for recovered ferrous metals: copper precipitation, remelt, detinning and secondary metal dealers.

Copper Precipitation
The light ferrous fraction is ideally suited to act as the precipitation agent in a leaching process for the beneficiation of copper ore utilized by copper refiners located in the western United States.

Remelt
Within this market there are three major industries all of which are primarily interested in the heavy fraction. These industries are: iron and steel producers, the ferroalloy industry and the iron and steel foundries. Since the scrap is used as a raw material in the production of new ferrous products, the scrap quality (i.e. level of residual alloys and other contaminants) is of great importance.

Detinning
The light fraction ("tin" cans) can be utilized by detinners as a source of high quality tin from chemical detinning operations. After detinning, the ferrous byproducts can be used by the copper precipitation or remelt markets.

Secondary Scrap Dealers
Scrap dealers often play a key role in the fluctuating ferrous scrap market. The scrap dealer is independent of any one customer and can therefore sell his scrap to the highest bidder in any of the above markets or in some other market (i.e. exporting).

As described above there are a variety of industries that

utilize scrap ferrous metals. <u>Market Locations for Recovered Materials</u> contains a listing of copper precipitation, detinning and steelmaking facilities in the U.S. Scrap dealers may be located in the local telephone directory (Yellow Pages).

Aluminum

Aluminum scrap recovered from MSW is comprised primarily of aluminum cans. It also contains rigid and flexible foil and other miscellaneous items.

Virtually all of the aluminum currently being recovered is recovered through source separation programs which are sponsored by either Reynolds Metal Company or the Aluminum Company of America.

Systems designed to recover aluminum from a mixed waste stream are still in the development and testing stages. The two technologies being developed are eddy current separation and electrostatic separation. The Hempstead, New York, Monroe County, New York and Dade County, Florida resource recovery facilities have either attempted or plan to incorporate aluminum recovery into their materials recovery systems. However, each of these facilities have experienced operational problems and as a result, significant operating data has not yet been collected.

The three prinicipal markets for recovered aluminum are: major aluminum producers, secondary aluminum smelters and aluminum scrap dealers.

<u>Major Aluminum Producers</u> - The Aluminum Company of America (AlCOA), Reynolds Metals, and Kaiser Aluminum are the three major producers in the United States and constitute the largest market for recovered aluminum. These companies operate aluminum recycling programs and have demonstrated an interest in purchasing refuse-derived aluminum scrap that is consistent with their specifications. In particular, Reynolds Metals, Inc. has contracts with resource recovery facilities in Hempstead, NY, Monroe County, NY and Dade County, FL.[1]

<u>Secondary Aluminum Smelters</u> - These firms buy scrap aluminum and process it into secondary aluminum ingot. Due to their lower energy requirements for production processes, secondary smelters can locate closer to the end-use consumption points. In contrast to this the

major aluminum producers must locate close to energy sources since the processing of bauxite is an energy intensive process. The net affect of this situation is that transportation costs to and from the major producers tend to be higher than for secondary smelters.

<u>Scrap Dealers</u> - A resource recovery plant that produces municipal aluminum scrap of a consistent quality would probably not sell it to a scrap dealer. However, an unprocessed nonferrous mix containing aluminum, brass, copper, and other metals could possibly be sold as scrap to a dealer. The price for such a mix would be low in comparison to recovered aluminum prices, however, no aluminum recovery system would be necessary, which means decreased processing costs.

As discussed above, the purchasers of scrap aluminum can be categorized into three groups. A listing of the major producers and secondary smelters, along with their locations, is provided in <u>Market Locations for Recovered Materials</u>. Local scrap dealers are often listed in the local telephone directory (yellow pages) under "recycled materials", "wastepaper dealers", or "scrap dealers".

<u>Glass</u>

Glass recovered from MSW is comprised of glass containers, flat glass and miscellaneous glass objects. Glass accounts for approximately 8 to 10 percent of MSW, with flint or clear glass accounting for about two thirds of the waste glass and the remaining third equally split between green and amber colored glass.

Virtually all glass recovered is by source separation programs. Mechanical recovery of glass from mixed waste at resource recovery facilities is still a developmental technology. Two promising technologies, froth flotation and optical sorting, are discussed in Section 4.

The largest market for color sorted glass cullet are glass container manufacturers. Recovered color mixed glass can also be utilized as a raw material in the manufacture of a number of products.

<u>Glass Container Manufacturers</u> - Waste glass is used in the manufacture of glass containers for the following

reasons: promotes melting at lower temperature, reduces fuel requirements, extends furnace life and decreases the time needed for melt to occur. Typically, cullet use is 10 to 20 percent of the glass batch, but it can be as high as 100 percent.

Historically, the scrap glass used is internally generated by the glass industry, since it is already color sorted, of known quality and free of contaminants. For glass recovered from a solid waste stream to be successfully marketed it must offer these same advantages to the glass container industry or other advantages, such as raw material cost reductions.

Other Products - There are number of secondary products that may be able to use cullet. Some of these potential markets are:

- o Road building materials such as slurry seal, glasphalt paving and glass beads for reflective paints.
- o Building materials such as bricks, foamed insulation, ceramic tiles, terrazzo tiles, sewer pipe, building blocks and aggregate.
- o Miscellaneous products such as glass-polymer composites, ground cover, trickling filters, jewelry, etc.

The quality of the glass that may be utilized in these products is considerably less stringent than for glass containers (i.e. mixed glass is useable and more contaminants can be tolerated). The use of scrap glass in these products is generally experimental, although it may be worth exploring the potential markets in a given area.

A listing of glass manufacturers located throughout the nation is in Market Locations for Recovered Materials. Local brick manufacturers and other construction materials manufacturers can be identified by looking in the local telephone directory (yellow pages) and other directories, which are discussed in the Sources of Information section.

Other Nonferrous Metals (exclusive of aluminum)

This group consists largely of red and white metals including: copper, brass, zinc, nonmagnetic stainless steel and alloys containing brass and lead.

Several technologies are undergoing testing to determine if they may be adapted to recover nonferrous metals other than aluminum. These technologies include jigging and heavy media separation. Other nonferrous recovery methods have been developed, but they are designed for aluminum recovery. It should be noted that recovery of other nonferrous metals has not been performed at a commercial scale resource recovery facility and there remains some uncertainty as to whether the recovery of these metals can be justified on economic grounds.

Secondary material processors (scrap dealers) are the primary market for other nonferrous metals. These markets should be contacted to determine if the metals are acceptable mixed together or if they must be separated, which is not yet technologically feasible. If the mixed material is acceptable it is probably valued by a composite price based on an assay of each delivery's yield.

Local scrap dealers, as discussed above, are the significant market for these metals from MSW and they can be located in the local telephone directory (Yellow Pages).

POTENTIAL ENERGY MARKETS
General

Solid, liquid and gaseous fuels, along with steam and electricity can all be recovered from municipal solid waste (MSW). The type of energy recovered is dependent upon the resource recovery technology utilized. However, not all of the technological options have developed to the point that may be considered for commercial scale facilities. The selection of an energy recovery system should be based not only on its technical feasibility, but also on the availability of ready markets for the products the system will provide. The energy products and markets to be discussed are:

- Solid fuel: fluff, powder and densified refuse derived fuel (RDF).
- Pyrolysis: gas and oil
- Methane gas
- Steam
- Electricity

This section will briefly define the energy products, the recovery methods, potential markets and specific purchasers.

Solid Fuels

The definition and recovery method of the three types of solid fuel will be separately discussed. The potential market and specific purchasers will be discussed together since these fuels are closely related in terms of marketability.

Fluff RDF - A refuse derived fuel (RDF) composed of the light combustible fraction contained in MSW. Particle size ranges from ¼ inch to 2 inches.

Fluff RDF is typically separated from the waste stream by using a combination of primary shredding, air classification, screening and secondary shredding. The separated fuel can then be stored, transported or fired. A more detailed discussion of this type of system is contained in the unit processes section. Examples of facilities that recover this type of RDF are the Ames, Iowa and St. Louis, Missouri resource recovery operations.

Powder RDF - This fuel is also composed of the light combustible fraction in MSW, however, its particle size is less than 0.15 millimeter. The only powder RDF developed to date is Combustion Equipment Associates Eco-Fuel II, a proprietary fuel. The recovery method is similar to that for fluff RDF. The differences are that the secondary shredder is eliminated, the light organic fraction is treated with an embrittling agent and the hardened material is then passed through a ball mill to produce the powder RDF.

Densified RDF - Densified RDF (d-RDF) is produced by pelletizing, extruding or briquetting fluff RDF or by adding a chemical binder to powder RDF and then densifying it into briquettes or pellets. See Chapter 4 for details.

The principal markets for RDF are large utility or industrial facilities, where the RDF would be used as a supplemental fuel. There has been some reluctance on the part of potential fuel markets to utilize RDF. This apprehension is primarily based on the heterogeneous nature of RDF and the deleterious effects RDF can have on the combustion system.

Additional information on specific markets for RDF is contained in the Sources of Information section.

Pyrolysis

Pyrolysis is the destructive distillation of organic matter carried out in an oxygen-free or in a low oxygen

environment. Products derived from the breakdown of organic matter vary depending on the particular process used, but the products can be classified into three major groups:
- o gases containing hydrogen, methane, and carbon monoxide.
- o a liquid or oil that includes acetic acid, methanol, acetone and other organics.
- o a solid "char" that is mainly carbon and ash, containing impurities, such as glass, rock and metal.

A variety of pyrolysis systems have been developed in an attempt to recover the energy present in MSW. The four principal systems, along with the type of fuel they recovery and the corporation primarily responsible for the development of the technology are provided below:
- o Medium temperature pyrolysis - Pyrolytic gas - Monsanto
- o Low temperature flash pyrolysis - Pyrolytic oil - Occidental Petroleum
- o Oxygen-fed slagging pyrolysis - Pyrolytic gas-Union Carbide
- o Air-fed slagging pyrolysis - Pyrolytic gas - Andco Torrax

Chapter 4 discusses these systems in detail. All of these systems have experienced significant problems and the only pyrolysis system that is currently available on a commercial basis is the Andco-Torrax system.

Since the end product of the Andco-Torrax system is steam, which is recovered in a waste heat boiler, the markets for it will be basically the same as those discussed in the energy section on steam.

Methane Gas

Methane gas or landfill gas (LFG) is generated by the natural decomposition of MSW in a landfill.

LFG is withdrawn from the landfill via a gas collection system and is then cleaned in one of two gas purification processes: minimum clean up, resulting in a medium-Btu gas (averaging 500-550 Btu/scf) and clean up to pipeline quality (1000 Btu/scf). A substantial capital investment in processing equipment is necessary to get higher Btu gas but this gas is worth substantially more than low Btu gas.

Separate markets exist for the two types of gas that are derived from the gas produced at landfills. The low Btu gas is suitable for use as boiler fuel or as a process fuel for industrial use. The high Btu gas, which has a high proportion of methane, is suitable for most activities now utilizing natural gas or for injection into a natural gas distribution system.[2]

Specific Markets for LFG include natural gas companies and industries utilizing natural gas in a manufacturing process or facilities using gas as a boiler fuel.

Steam (or Hot Water)

Steam is produced from MSW by firing it in an incinerator and then recovering the heat in a boiler. A wide range of system configurations are possible, and are discussed in Section 4. Steam can also be produced from MSW by burning any of the solid (RDF) or when burning pyrolysis or firing LFG.

Possible uses of steam produced from MSW are the same as the uses for steam produced from any other fuel. These uses include: process heating, driving steam turbines and process equipment, space heating and cooling, air compression and cleanup. The three principal markets that coincide with these uses are: district heating and cooling systems, industrial plants, and steam electric power plants.

Electricity

Electrical generating equipment (typically a steam turbine) may be added to a steam producing system. See the technology section for more detailed information.

There are two principal markets for recovered electricity: utilities and specific customers, such as large manufacturing facilities. Rates for electricity sold to specific customers are generally substantially higher than the wholesale rates a utility would pay. However, the higher rates obtainable from a specific customer may be more than offset by the additional costs associated with supplying electricity to him. These costs include: the need for power lines to the customer, backup equipment to provide the customer with power in the event of equipment failure or required maintenance, and the potential institutional barriers in the form of public bonding constraints associated with the sale of energy to private customers.[3] However, before selecting one of these two markets, a thorough analysis of both markets should be completed to determine which market provides the greatest benefit to the community.

A recent article discusses the federal regulations that have been passed that should provide a reliable, long-term market with reasonable prices for many recovery facilities producing electricity.[4] As this article points out, these regulations, issued by the Federal Energy Regulatory Commission (FERC), provide that:

- o Utilities must purchase electricity from qualifying facilities at an appropriate rate (to be determined by state regulatory agencies, but incorporating cost elements of capital offset and fuel replacement).

- o Qualifying facilities that produce and sell electric energy, are exempt from federal and state regulations pertaining to electric utilities.

- o Utilities must provide qualifying facilities with electric energy and other certain types of services which may be requested by a qualifying facility to supplement or back up those facilities own generation.

FERC issued these regulations to implement the Public Utility Regulatory Policies Act (PURPA). For a resource recovery project to qualify for the benefits of these regulations it must have a capacity of less than 80 megawatts or be designated as a cogenerating facility. By March 1981, state agencies must comply with the PURPA regulations set forth by FERC. A detailed discussion of PURPA is included in Chapter 9.

PRODUCT SPECIFICATIONS
General

Specifications for recovered materials and energy from MSW are in a state of flux. Many of these recovered products are unknown quantities, often with little or no data available concerning their use. Recently, several organizations have developed tentative specifications while others, such as the aluminum industry, have published detailed specifications delineating percentages of contaminants and several grades of the recovered material. The American Society for Testing and Materials (ASTM) has formed a committee (E-38) whose purpose is to develop specifications and test methods for materials and energy resources recoverable from MSW. Although the majority of these specifications are not finalized, a combination of the specifications available from the various industrial institutes and ASTM can serve as a basis for evaluating recovered materials and energy.

Paper

Specifications for newspaper recovered through source separation programs have been developed by the Paper Stock Institute (PSI) and are similar to those used by the National Association of Recycling Industries (NARI). Appendix B contains those PSI specifications that are applicable to newsprint recovered via source separation.

Target specifications for recovering paper and old corrugated boxes from a mixed MSW stream have been developed and are presented in Appendix B. Two problems are associated with the actual use of these specifications. First, it is not certain that these specifications can be met utilizing any of the recovery technologies presently available. Second, at resource recovery facilities the paper fraction of the mixed MSW is usually left in the processed waste because of its fuel value and hence it is not available for recovery. Another problem associated with recovering paper from mixed MSW is the absence of established markets for such a product. With no markets available, it is difficult to generate interest in developing specifications for paper recovered from mixed MSW.[5]

Ferrous Metals

Traditionally, the specifications used for scrap ferrous metals have been issued by the Institute of Scrap Iron and Steel (ISIS), Inc. These are contained in the ISIS bulletin <u>Specification for Iron and Steel Scrap</u>. A listing of the scrap iron grades that are appropriate for municipal ferrous scrap is contained in Table 3-1. During 1980 ASTM adopted a specification (E702-79) and a test methodology (E 701-80) for municipal ferrous scrap. These documents are specific to ferrous metals recovered from MSW and are available from ASTM.

Aluminum

Standard specifications for municipal aluminum scrap have

been deveoped by Reynolds Metals and Alcoa and are presented in Appendix B. These specifications are divided into grades based on the chemical composition of the aluminum scrap, which determines its economic value.

Glass

A standard specification for waste glass for use in the manufacture of glass containers has been developed by the ASTM. This standard is designated E708-79 and is similar to a tentative specification developed by the Glass Packaging Institute in 1976. A copy of each of these specifications is reproduced in Appendix B. These specifications set stringent guidelines for the amount of color mixing and contamination levels permitted. Until glass recovery systems are tested and operational on a full scale level, it is uncertain whether these specifications can be met on a commercial production scale.

Nonferrous Metals (exclusive of aluminum)

Specifications for nonferrous metals are more qualitative than those for other materials. This is a result of the nonferrous metals value being established on an "assay" basis after it has been received by the purchaser. In one situation where nonferrous metals were to be recovered and sold, the only specification was that the material be clean and free of residue such as paper, trash or dirt.[6] The presence of such contaminants would result in a less than optimum price for the recovered metals.

RDF

The ASTM has designated a subcommittee (E-38.01) to develop specifications for refuse derived fuels (RDF). This committee expects to make ten standards available in 1983 including one dealing with what the ASTM calls RDF3.[7] RDF3 is a processed RDF composed of the light organic fraction of MSW. When these standards become available they will be listed in Part 41 of the Annual Book of ASTM Standards. Until that time, several physical and chemical factors may be employed in evaluating the marketability of an RDF:

> Moisture Content - An RDF should have as low a moisture content as possible prior to being fed into the

combustion unit, since the combustion efficiency of the unit is reduced as the moisture content is increased. Typically, an RDF should have a moisture content of 25 percent or less.

Ash Content - The residue remaining after combustion is ash. The fuel firing system and furnace walls can be adversely affected by ash slagging and erosion. Thus, the composition and quantity of the ash are critical to the efficient operation of the combustion processes. In additon, ash disposal is a consideration, since material handling and disposal costs rise as the ash content increases.

Particle Size - The type of combustion unit determines the desirable particle size. Particle size can range from 0.15 millimeter for Eco-Fuel II, projected for use in full suspension units, to raw MSW (no size reduction) used in mass fired incinerators. Processing costs generally increase as the particle size decreases.

Energy Yield - The energy yield of RDF is typically measured as Btu per pound as received. The percentage of moisture content has a significant impact on energy yield and therefore must be kept low enough (25 percent or less) to permit efficient operation of the combustion unit.

Methane Gas

To establish the specifications for high Btu gas, the purchaser (usually the local utility) should be contacted to determine their requirements. Generally, the composition of the high Btu gas is 90 percent methane with an energy yield of over 700 Btu/cu ft. These characteristics should be suitable for the utility, but they should be confirmed along with requirements regarding moisture content and contaminants.

Specifications for low Btu gas have not been developed because this gas is usually limited to on-site uses and is not normally suitable for injection into a pipeline or for other external use.

Steam

Specifications for steam are dependent upon the purchaser's needs. Specific users may have different requirements regarding the temperature and pressure of the steam. These requirements should be known well in advance so that the recovery system may be designed to meet these needs. An important consideration is that the specifications of the purchaser must be within the temperature and pressure limits that are associated with producing steam when MSW is the sole or primary fuel. To minimize slagging and corrosion of the boiler tubes, steam is usually produced at 600 psi or less, when MSW is used as a fuel. Temperatures can generally range from 250°F to over 1,000°F depending on whether the steam is for electric power generation (high temperature) or steam distribution systems (lower temperatures). Condensate return is also a factor in the specification of a steam system. Return of condensate to the boiler is usually desirable because it decreases the quantity of feedwater treatment chemicals required. The condensate, however, must comply with minimum quality specifications for reuse in the steam boilers.

ESTIMATING REVENUES
General

The value of recovered energy and material from a resource recovery facility is set by market forces just as for any other commodity. The price for each product is therefore determined by the price of other competing commodities.

The marketability of a material or energy product is influenced by a number of factors including:

- o the demand for such a product;
- o its quality, including the degree to which it meets established specifications;
- o the cost of delivering to a customer;
- o the price of an alternative material or energy source; and
- o any additional manufacturing costs due to using a recovered rather than virgin product.

Inadequacies or uncertainties in any of these factors can impair the marketability and reduce the value of a recovered resource.

Ferrous Metals

The price of light and heavy ferrous has been historically tied to the price of the No. 2 Bundle and No. 1 Heavy

Melting Steel (see Table 3-1) respectively as published in Iron Age Magazine. The price for light ferrous in typically 50 to 60 percent of the price for a No. 2 Bundle; for heavy ferrous 80 to 100 percent of the price for heavy melting steel (HMS).

Historically, prices and demand for ferrous scrap have fluctuated widely. Such fluctuations have resulted in some ferrous recovery systems operating at a loss, and others unable to market their products at all. A resource recovery facility can attempt to isolate itself from market fluctuations by obtaining a long term contact for ferrous sales.

TABLE 3-1

INSTITUTE OF SCRAP IRON AND STEEL (ISIS), INC.
PARTIAL LISTING OF SCRAP IRON GRADES

ISIS Code Number	Classification Description
200	No. 1 Heavy Melting Steel: Wrought iron and/or steel scrap 1/4 inch and over in thickness. Individual pieces not over 60 x 24 inches prepared in a manner to insure compact charging.
209	No. 2 Bundles: Old black and galvanized steel sheet scrap, hydraulically compressed to charging box size and weighing not less than 75 pounds per ft^3. May not include tin or lead - contact material or vitreous enameled material.
213	Shredded Tin Cans for Remelting: Shredded steel cans tin coated or tin free, may include aluminum tops but must be free of aluminum cans, nonferrous metals except those used in can construction and nonmetallics of any kind.
214	No. 3 Bundles: Old sheet steel compressed to charging box size and weighing not less than 75 pounds per ft^3. May include all coated ferrous scrap not suitable for inclusion in No. 2 Bundles.
215	Incinerator Bundles: Tin can scrap, compressed to charging box size and weighing not less than 75 pounds per ft^3. Processed through a recognized garbage incinerator.

Source: ISIS Bulletin Specification for Iron and Steel Scrap, 1975.

Other Non-Ferrous (ONF)

The value of mixed non-ferrous metals is a composite price based on an assay of each delivery. A typical composition for this material is given in Table 3-2.

TABLE 3-2

Typical ONF Composition

Metal	Percent by Weight (Total ONF Metals)
Copper	56.0
Zinc	35.0
Tin	1.0
Lead	8.0
	100.0

The preseparation value of the ONF fraction can be estimated from average scrap prices for each component (excluding tin which is unrecoverable), obtained from the American Metal Market. Post-separation values, i.e., reclamation of ONF after incineration from the bottom ash are generally 50 to 75 percent of preseparation, due to losses in the incineration process.

Aluminum

According to Reynolds Aluminum, the pricing of scrap aluminum recovered from MSW is pegged to the price of old aluminum sheet and casting, which is published in *American Metal Market*.[8] For example, the price of MSW aluminum meeting the specifications of Grade A aluminum scrap (see Appendix B) is generally approximately 90% of the old aluminum sheet and casting price. Using January 1981 prices, a ton of Grade A MSW aluminum would be valued at between $450 to $666 per ton depending on the location of the facility. Grade B MSW aluminum (see Appendix B) is usually valued at approximately 20% less than Grade A MSW aluminum.

Paper

Two paper commodities are generally recovered from solid waste: newsprint and corrugated. The value of each depends on market conditions and specifications. For newsprint, the price is generally a percentage of the price of No. 1 newsprint (Grade 6) as quoted in the Official Board Market (Yellow Sheet); for corrugated the reference commodity is corrugated containers (Grade 11).

Glass

    Glass cullet value can be estimated:
- by contacting a local glass manufacturer who uses cullet on a constant basis;
- by calculating its materials substitution value (Data on various mixes used in manufactoring glass can be obtained using the <u>Minerals Yearbook</u> published by the US Department of the Interior); or
- by using the price paid for glass recovered through source separation programs.

Typical Estimated Materials Products Revenues

Table 3-3 presents a summary of materials revenues from a typical MSW composition, assumed recovery efficiencies, and estimated current floor and market prices.

Recovered Energy Products

Just as for recovered materials products, the price of energy products are determined by considering the price of other competing products. In general, a more refined energy product can replace a less refined, if prices are competitive: the reverse is generally not possible. Thus high-Btu RDF (a highly refined energy product) might replace coal (a less refined product) in an industrial boiler; MSW could not replace natural gas.

In marketing recovered energy products, it must be remembered that there are several factors which make such products less desireable than convention energy sources to potential purchasers. These factors include:
- uncertainty about quality of the energy product;
- uncertainty about reliability of supply;
- the need for modification of equipment and/or operating procedures by the energy purchaser; and
- reluctance to alter long-term relationships with traditional energy suppliers.

Because of such considerations, it must be expected that recovered energy products will have to be marketed at a discount relative to conventional energy sources.

TABLE 3-3

ESTIMATED MATERIALS REVENUES[1]

Materials Product	% in MSW	Recovery Efficiency (%)	Recovered, % of MSW	Market Value ($ per ton)[2]		$/Input Ton of MSW (Revenue)	
				Exchange	Floor	Exchange	Floor
Ferrous							
Light	8.7	97	8.44	$25	$20	$2.11	$1.69
Heavy	0.8	83	0.66	$40	$20	$0.26	$0.13
Other							
Non-Ferrous	0.2	57	0.11	$250	$200	$0.29	$0.23
Aluminum	0.5	57	0.29	$700	$600	$2.00	$1.71
Glass	6.6	73[3]	4.82	$20	$15	$0.96	$0.72
Paper							
News	8.0	30	2.40	$20	$12	$0.48	$0.29
Corrugated	5.0	50	2.50	$18	$10	$0.45	$0.25
					Total	$6.55	$5.02

[1] These are gross revenues, not including transportation.

[2] Estimates of reasonable expectations for floor, or lowest anticipated, sales prices are given to provide guidance on the minimum anticipated revenues.

[3] Efficiency for fines recovery.

Typical Estimated Energy Products Revenues

Table 3-4 presents a summary of energy product revenues, assuming MSW with a heating value of 4500 Btu per lb, typical recovery and conversion efficiencies, and estimated prices for competing conventional energy products. This table does not include preparation or transportation which may add significant costs to the process of obtaining revenues. For instance, light ferrous may have to be compressed, baled and delivered to the user and these costs must be subtracted from the expected revenues.

MARKET SECUREMENT

The purpose of collecting information concerning potential purchasers is to provide a base for a market survey. Once a list of potential material and energy purchasers has been compiled, a questionnaire should be prepared and sent out to potential users to determine if they are interested in the recovered products. Typically, a questionnaire should request information on the type and amount of recovered product, where the potential user is located (the office to where the questionnaire is sent may not be located at the facility where the products are required), type of transportation required, contractual requirements (time period, price formula, specifications, etc) and a space for comments or additional information. Examples of materials and energy questionnaires are provided in Appendix A.

The results of the questionnaire should be incorporated into the technology selection phase of the planning process. This is imperative, since the customers specifications for the recovered products dictate the type of recovery process(es) to be used, and how they are to be operated to meet the specifications. Prior to the selection of a resource recovery system firm purchase commitments should be obtained for the materials and energy to be recovered. Initially, a letter of interest may be received from a potential buyer. This type of letter outlines, in general terms, that the potential user is interested in negotiating some form of an aggeement to purchase one or more of the recovered products.

Appendix A contains an example of a letter of interest. A letter of interest does not provide a market guarantee, since it only discusses a willingness to consider future purchase of recovered products.

TABLE 3-4

ESTIMATED ENERGY PRODUCTS REVENUES

Energy Product	Fuel Type	Heating Value Btu/unit	Alternative Costs					Estimated Sale Price $/unit	Conversion Rate	Energy output Per input ton Unit/ton	Estimated Revenues $/input ton
			Price $/unit	$/MMBtu	Conversion Efficiency	Cost of Steam $/MMBtu				
Steam	No. 2 Oil	140,000/gal	$1.00/gal	$7.14	80%	$8.93	$7.00/MMBtu	2.5 lbs steam/lb MSW	5.0 MMBtu	$35.00
	No. 6 Oil	150,000/gal	$30/barrel	4.76	80%	5.95	4.60/MMBtu	2.5 lbs steam/lb MSW	5.0 MMBtu	23.00
	Coal	13,000/lb	$60/ton	2.31	75%	3.08	2.25/MMBtu	2.5 lbs steam/lb MSW	5.0 MMBtu	11.25
	Natural gas	1020/CF	$2.50/1000CF	2.45	85%	2.88	2.00/MMBtu	2.5 lbs steam/lb MSW	5.0 MMBtu	10.00
Gas	No. 2 Oil	140,000/gal	$1.00/gal	$7.14	---	---	6.00/MMBtu	0.75 Btu/Btu input	6.75 MMBtu	$40.50
	No. 6 Oil	150,000/gal	$30/barrel	4.76	---	---	3.50/MMBtu	0.75 Btu/Btu input	6.75 MMBtu	23.63
	Coal	13,000/lb	$60/ton	2.31	---	---	1.75/MMBtu	0.75 Btu/Btu input	6.75 MMBtu	11.81
	Natural gas	1020/CF	$2.50/1000CF	2.45	---	---	1.80/MMBtu	0.75 Btu/Btu input	6.75 MMBtu	12.15
Solid Fuel	Coal	13,000/lb	$60/ton	$2.31	---	---	1.75/MMBtu	0.85 Btu/Btu input	7.65 MMBtu	13.39
Electricity (in-house)	Electricity	---	$0.05/kwh	---	---	---	$0.05/kwh	13,000 Btu/kwh	692 kwh	34.60
(utility sales)	Electricity	---	$0.03/kwh	---	---	---	0.03/kwh	13,000 Btu/kwh	692 kwh	20.76

The letter of intent (LOI) is a more formal and specific document. A LOI is binding and provides a guarantee to buy recovered material or energy, since it is a contract between buyer and seller in the form of an advance commitment to buy certain products under specified conditions.[9] This type of arrangement provides the organization responsible for the project with a firm revenue base for planning the facility and selecting the appropriate technology.

To ensure that the necessary commitment has been received, the following components should be included in an LOI:

o length of commitment

o delivery guarantee

o quantity of material

o technical specifications

o termination

o pricing

The length of commitment preferably should be for the first five operating years of the facility, while the delivery guarantee prevents the seller from diverting recovered products to "spot" markets having higher prices. The quantity of material to be sold may be expressed as a range for the first year, to account for the uncertainty of the waste composition and should be subject to adjustment after the first year. The conditions under which the product(s) will be accepted or rejected should be described in the technical specifications. A section describing the conditions for terminating the agreement should be included in an LOI. This clause may be provided to release the purchaser in the event of an inordinate delay in the implementation of the project.

An important component of an LOI is its establishment of price. A number of methods may be utilized to determine the price of the recovered product, including: a fixed price, percentage of a commodity or scrap quotation, and the price of similar materials. Irrespective of which pricing method is used, a floor or minimum price should be included, since without it reliable forecasts can not be formulated.

The use of an LOI by publicly-owned facilities may not be permitted, since these facilties are generally required to employ public bidding when selling recovered materials or energy. An innovative solution, used by the Tennessee Valley Authority and in Allegany County, Pennsylvania was to secure a letter of intent to bid (the above discussed LOIs were letters of intent to buy). An LOI to bid is essentially an agreement by the potential purchaser to submit a response to a future invitation to bid for the purchase of recovered material or energy. The terms and conditions describing the purchase can be very similar to those of an LOI to buy, and thus permit serious financial planning to continue for a publicly-owned facility. A sample LOI to bid is contained in Appendix A.

Once the LOIs for the materials and energy have been received the market assessment for the project is virtually complete. Formal purchase contracts may be developed, but the LOIs are usually sufficiently binding to protect the long term interest of the facility. Specific purchase orders, which will be issued by the purchasers during the operational phase of the project, are the remaining marketing agreements to be developed.

SOURCES OF INFORMATION
General

A list of potential materials and energy purchasers can be developed through consultation with a number of information sources.

Materials Purchasers

For potential materials purchasers, the following sources should prove useful:

- Market Locations for Recovered Materials Stephen E. Howard, U.S. Environmental Proection Agency (SW-518) August 1976.

- Industrial directories or information available from the various trade associations such as:

    - National Association of Recycling Industries

    - Aluminum Association

    - Glass Packaging Institute

    - Institute of Scrap Iron and Steel

- The area telephone directories, particularly the yellow pages, which are frequently an excellent source of information on local scrap dealers and the types of materials that are of interest to these dealers.

- o Standard Industrial Code Classification information from the US Department of Commerce.

- o State and local chamber of commerce or equivalent organizations.

- o Reports and other informational sources on resource recovery planning and implementation work done within the region in which the project is planned.

Energy Users

For the potential energy users, the following sources should prove useful:

- o The local, state, and/or regional agency responsible for monitoring air pollution levels and permit requirements of facilities with large electrical generating equipment.

- o "Study of the Feasibility of Federal Procurement of Fuels Produced from Solid Wastes", Arthur D. Little, Inc. for U.S. Environmental Protection Angency, July 1975.

- o U.S. Department of Commerce records on identification and classification of major industrial and commercial employers in the region.

- o Information and knowledge concerning the involvement of potential energy users in past and present resource recovery efforts in the region.

REFERENCES

1. Telephone conversation on December 18, 1980 between Mr. Gilbert Boucier, Manager of Environmental Planning Recycling and Reclamation Technology for Reynolds Metals Company and Mr. Robert E. Bolton (MPI).

2. This section is adapted from "Landfill Methane: 23 sites are developing recovery programs" by Robert P. Stearns in Solid Waste Management, June 1980.

3. Brown & Caldwell, Solid Waste Management Resource Recovery Feasiblity Study for Hillsborough County Florida, December 1, 1979, p 3-16.

4. Aldrich, R.H. and Rene Rofe, "Resource Recovery: An Investment Opportunity for the '80s," NCRR Bulletin (December 1980).

5. Telephone conversation on January 12, 1981 between Mr. Ken McDaniel of Miami Paper Co., former chairman of the ASTM subcommittee on Paper and Paperboard (E-38.04) and Mr. Robert E. Bolton (MPI).

6. National Center for Resource Recovery, Inc., New Orleans <u>Resource Recovery Facility Implementation Study Equipment, Economics, Environment</u>, September 1977, p. 188.

7. Telephone conversation on January 12, 1981 between Dr. John Love, chairman of the ASTM subcommitte on Energy (E-38.01) and Mr. Robert E. Bolton (MPI).

8. Telephone conversation on December 18, 1980 between Mr. Gilbert Boucier of Reynolds Aluminum and Mr. Robert E. Bolton (MPI).

9. Vesilind, P.A. and Warner, D. "Managing the risk aspects through planning and market agreements" <u>Solid Waste Management</u>, 1980 Sanitation Industry Yearbook.

4

Unit Processes of Resource Recovery

GENERAL

This chapter presents the unit processes of resource recovery and is divided into four sections:
- o Preprocessing
- o Energy recovery
- o Materials recovery
- o Materials handling

PREPROCESSING

Preprocessing of solid wastes prior to incineration is desirable to yield a waste stream of greater homogeneity and permit recovery of materials such as aluminum, glass, and ferrous metals. This chapter presents the preprocessing activities which are normally utilized at a resource recovery facility, as follows:
- o Weighing
- o Receiving and Storage
- o Trommel Screening
- o Shredding
- o Air Classification

Weighing Facilities

Weighing of incoming waste is necessary to provide accurate information on the quantity of wastes received and help regulate the operation of the facility. In addition, fluctuations and varying trends in the quantity and types of wastes being brought to the facility can be ascertained. Other reasons why weighing waste is important are to:

- o establish equitable fees for processing the wastes;
- o optimize fuel consumption of collection vehicles by checking the compaction efficiency of trucks;
- o establish more efficient route collection systems;
- o protect against truck overloads;
- o prevent private haulers from jamming trucks (this is possible if fees are based on yardage, and not weight).

There are three main types of scales used for weighing the wastes: beam weight, load cell and a combination mechanical/electronics scale. The beam weight scale operates a system of levers and requires excavation of a pit. A typical beam weight scale[1] is shown in Figure 4-1. The load cell type requires less excavation but requires protection from shock and side loads. It has at least four load cells at its corners and electronically determines the truck weight by the load from each cell. The combination scale has a mechanical beam arrangement which transfers the weight to a single cell, usually located beneath the center of the scale platform.

In selecting a scale, a 50 ton capacity is usually adequate. Platforms 10 feet by 34 feet in length are sufficient for weighing packer trucks. However, if tractor-trailer transfer vehicles will be bringing wastes to the facility a 50-foot platform length is required. Adequate provisions should be made for signal lights, curbing, alarms and automated recording devices. In addition, snow, ice and debris should be cleared from the platform. Periodic inspection of the scale to assure clearances to 1.0 percent should be conducted and include: [2]

- o check of the indicated weight as a heavy load is moved from the front to the back of the scale;
- o observation of the dial for irregularities; and
- o use of test weights.

The information which should be recorded when trucks are weighed include:

- o date and time

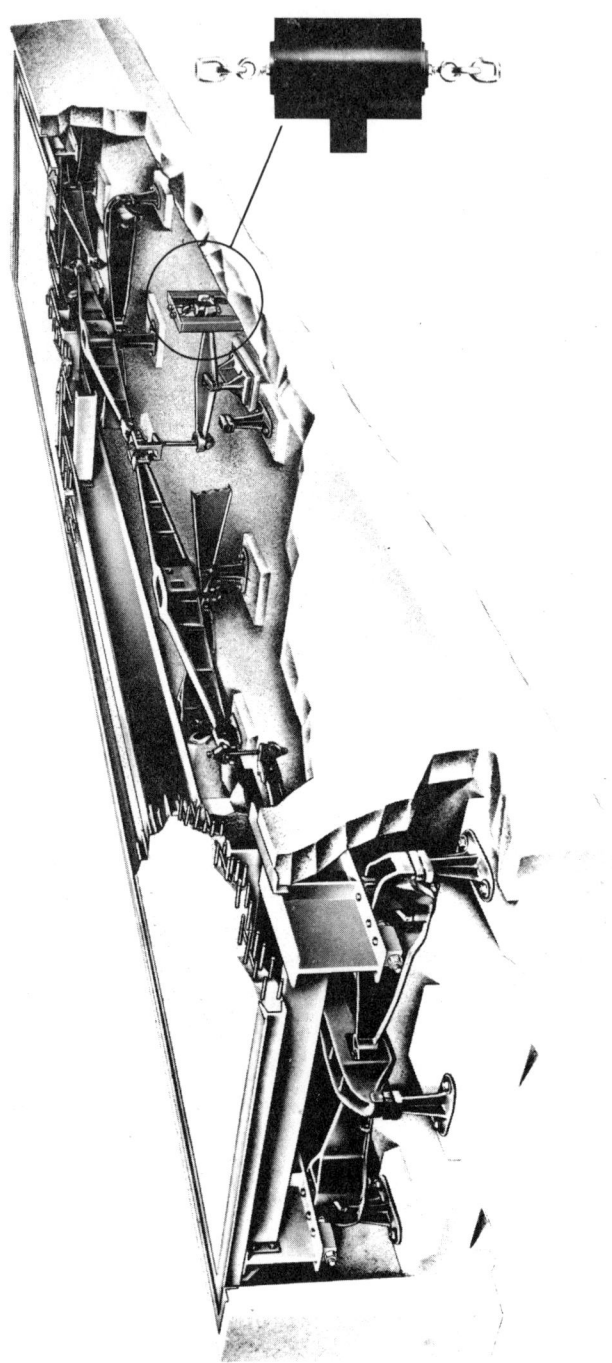

Figure 4-1
Beam Weight Scale
(Fairbanks Division, Colt Industries)

o   vehicle identification

o   tare weight

o   gross weight

o   net weight

Automated systems which print this data onto cards are available. A sample card with this type of information is shown in Figure 4-2.[3] Sample summary reports that this type of system can generate are provided in Figure 4-3.

The scale system must be adequate to handle the maximum expected number of vehicles without excessive delays or backups. With an automated system, approximately one truck can be weighed each minute, or 60 trucks per hour per scale. Since trucks arrive in a random pattern a simple queuing (waiting) time analysis is helpful. Example 4-1 illustrates such an analysis.

EXAMPLE 4-1

Consider a resource recovery facility receiving an average of 45 vehicles per hour, equipped with one scale.

1. Calculate average length of waiting line.

   A = average arrival rate = 45/hr
   S = average service rate = 60/hr
   N = number of service stations = 1

   Average Queue = $\dfrac{A^2}{S(S-A)}$ = $\dfrac{45^2}{60(15)}$ = 2.25 vehicles

2. Calculate average waiting time.

   Average waiting time = $\dfrac{A}{S(S-A)}$ = $\dfrac{45}{60(15)}$ = 0.05 hours = 3 mins

3. If the arrival rate increases to 55 vehicles per hour during peaks, calculate average length of queue and average waiting time.

   Average Queue = $\dfrac{55^2}{60(5)}$ = 10.08 vehicles

   Average Waiting Time = $\dfrac{55}{60(5)}$ = 0.18 hours = 11 mins

Receiving and Storage

Receiving/storage refers to that portion of the resource recovery facility which receives incoming refuse vehicles,

54 Energy and Resource Recovery from Waste

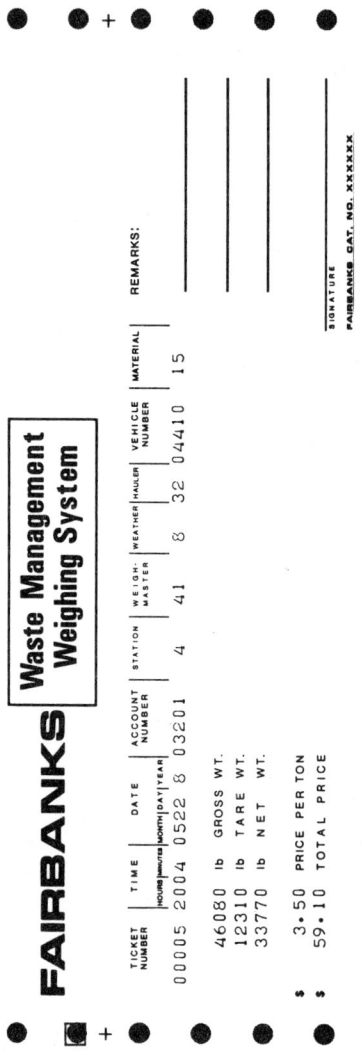

Figure 4-2
Sample Card
(Fairbanks Division, Colt Industries)

Unit Processes of Resource Recovery 55

```
                FAIRBANKS SUMMARY REPORT 1    PAGE 21
                LISTED BY TRANSACTION


PRINTED AT STATION 6       19:59   01/15/8

DATE    TIME    ACCOUNT    VEHICLE    NET WT    TOTAL    MATERIAL
                NUMBER     NUMBER     POUNDS    PRICE
01/15   15:33   46023      66000      33920     43.04    17
01/15   15:33   12023      2743       27220     77.20    19
01/15   15:34   12023      2407       27460     46.72    01
01/15   15:36   63023      1350       10460     16.26    04
01/15   15:37   04323      7600       28160     68.12    07
01/15   15:37   07766      3200       29340     72.41    08
01/15   15:38   00466      21200      34340     32.83    17
```

```
                FAIRBANKS SUMMARY REPORT 2    PAGE 1
                LISTED BY ACCOUNT  NUMBER


PRINTED AT STATION 6       20:13   01/15/8


ACCOUNT    VEHICLE    NUMBER     NET WT    TOTAL
NUMBER     NUMBER     OF LOADS   POUNDS    PRICE
00001      1200       8          160660    281.50
           76070      9          199940    349.89
           99470      8          175440    307.02

TOTALS FOR ACCOUNT    25         536240    938.41

00011      2031       11         238820    417.94

TOTALS FOR ACCOUNT    11         238820    417.94

00053      5010       13         255420    446.96
           5000       5          99840     174.72

TOTALS FOR ACCOUNT    18         355260    621.68

00100      1090       7          153020    267.78

TOTALS FOR ACCOUNT    7          153020    267.78
```

```
                FAIRBANKS SUMMARY REPORT 3
                LISTED BY MATERIAL TYPE

PRINTED AT STATION 6       20:33   01/15/8


MATERIAL    NUMBER      NET WT     TOTAL
TYPE        OF LOADS    POUNDS     PRICE

0           4           75080      87.84
1           19          295840     510.32
2           18          372860     850.12
3           51          604540     1,713.87
4           20          317820     1,077.41
5           23          464140     1,831.03
6           42          507480     2,278.59
```

```
SYSTEM STATUS AT 17:45 01/15/8   STATION 6
00929 LAST TICKET NUMBER
0926 LOADS IN MEMORY

MAT    PRICE      ACCUMULATION
       ($)        (TONS)
00     002.34     0037.54
01     003.45     0067.49
02     004.56     0086.46
03     005.67     0064.55
04     006.78     0040.23
05     007.89     0113.62
06     008.98     0120.01
07     009.87     0070.74
08     008.76     0129.16
09     007.65     0046.71
10     006.54     0091.47
11     005.43     0165.80
12     004.32     0173.04
13     003.21     0094.78
14     002.10     0183.84
15     004.01     0086.10
16     005.00     0101.17
17     009.00     0413.94
18     009.99     0067.71
19     009.50     0230.96
```

Figure 4-3
Summary Reports
(Fairbanks Division, Colt Industries)

provides space and time for them to unload, and provides storage for raw refuse before processing.

Receiving/storage plays an important function in buffering the processing sections of the facility from the sharp variations in the rate of incoming refuse. Most resource recovery facilities operate continuously, while refuse is generally collected only five days a week, one shift a day. Example 4-2 illustrates the impact of this variation.

EXAMPLE 4-2
Consider a 1000 tpd resource recovery facility. Assume all MSW is received in 20 cu yd. compactor trucks, at 500 lbs/cu yd.

$$\text{Truck capacity} = \frac{20 \text{ cu yd} \times 500 \text{ lb/cu yd}}{2000 \text{ lb/ton}} = 5 \text{ ton/truck}$$

$$\text{Nominal processing rate} = \frac{1000}{24} = 42 \text{ tph}$$

$$42/5 = \frac{8.4 \text{ truckloads}}{\text{hr}}$$

But, collection proceeds only 5 days/wk, 1 shift/day whereas the facility processes waste 24 hours a day, 7 days a week. In addition, only about 6 hours/shift are actually used due to lunch, cleanup, etc. Therefore:

$$\text{Actual receiving rate} = \frac{1000 \times 7 \text{ days}}{5 \text{ days} \times 6 \text{ hr/day}} = 233 \text{ tph}$$

$$\frac{233 \text{ tph}}{5 \text{ ton/truck}} = \frac{47 \text{ truckloads}}{\text{hr}}$$

Furthermore, there is an hourly peaking factor of about 150%, plus a seasonal peaking factor of 25%.

$$\text{Peak receiving rate} = 233 \times 1.5 \times 1.25 = 437 \text{ tph}$$

$$436/5 = 88 \text{ truckloads/hr}$$

The storage capacity is a function of process design and depends to some extent on the storage provided at other points in the facility. For facilities processing continuously, and without other storage, generally three day capacity must be provided to handle long weekends.

There are four general types of receiving/storage facilities for as-received refuse:
- o Pit and Crane;
- o Tipping floor;
- o Depressed tipping floor; and
- o Live bottom.

Unit Processes of Resource Recovery   57

<u>Pit and Crane</u> - Shown in Figures 4-4 and 4-5, this is the oldest type of system and is commonly used in mass burning incineration systems. It has the advantage of proven technology and requiring relatively little area. Its disadvantages include high construction cost, high maintenance costs on the crane, difficulty in cleaning the pit, and difficulty in controlling fires in the pit.

Figure 4-4
Pit and Crane
(Wheelabrator-Frye Inc.)

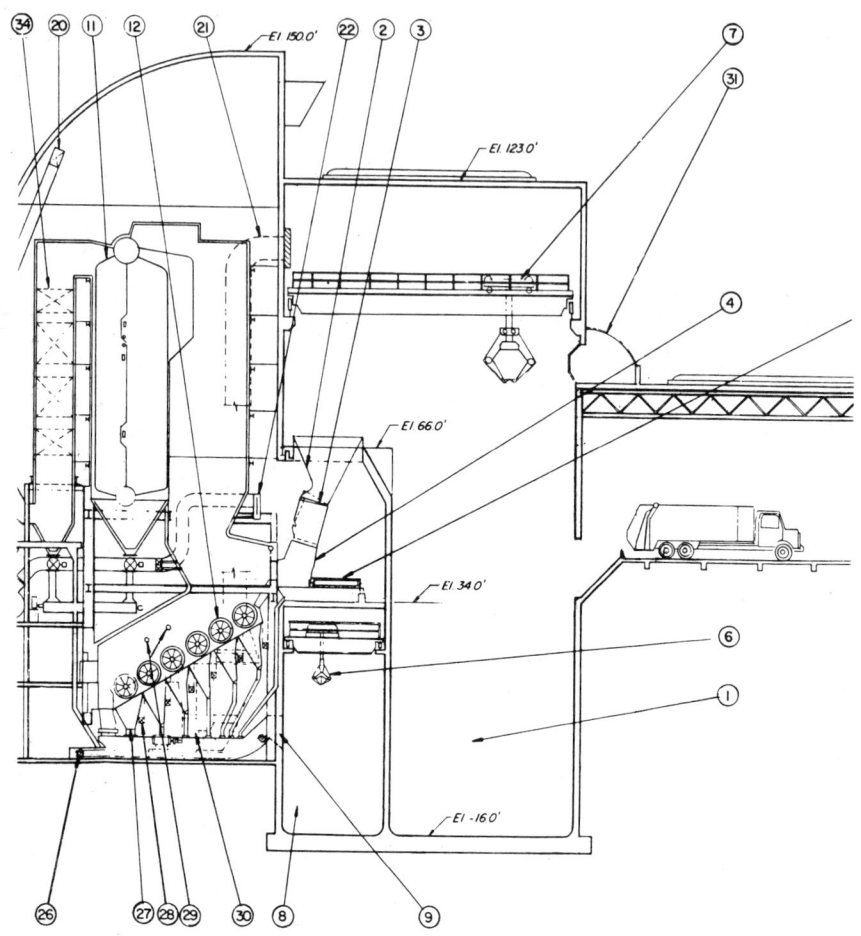

Figure 4-5
Pit and Crane
(Grumman Ecosystems Corp.)

Tipping Floor - Shown in Figure 4-6, this is a newer development and is often used where front end processing is intended. Its advantages include the ability to clean the floor regularly, the ability to presort the refuse to eliminate unprocessable and hazardous items, and that refuse handling is accomplished with proven, reliable equipment (wheeled and skid loaders). Its disadvantages include the need for a large, unobstructed floor space, and its unadaptability for more than a few hours storage.

Unit Processes of Resource Recovery   59

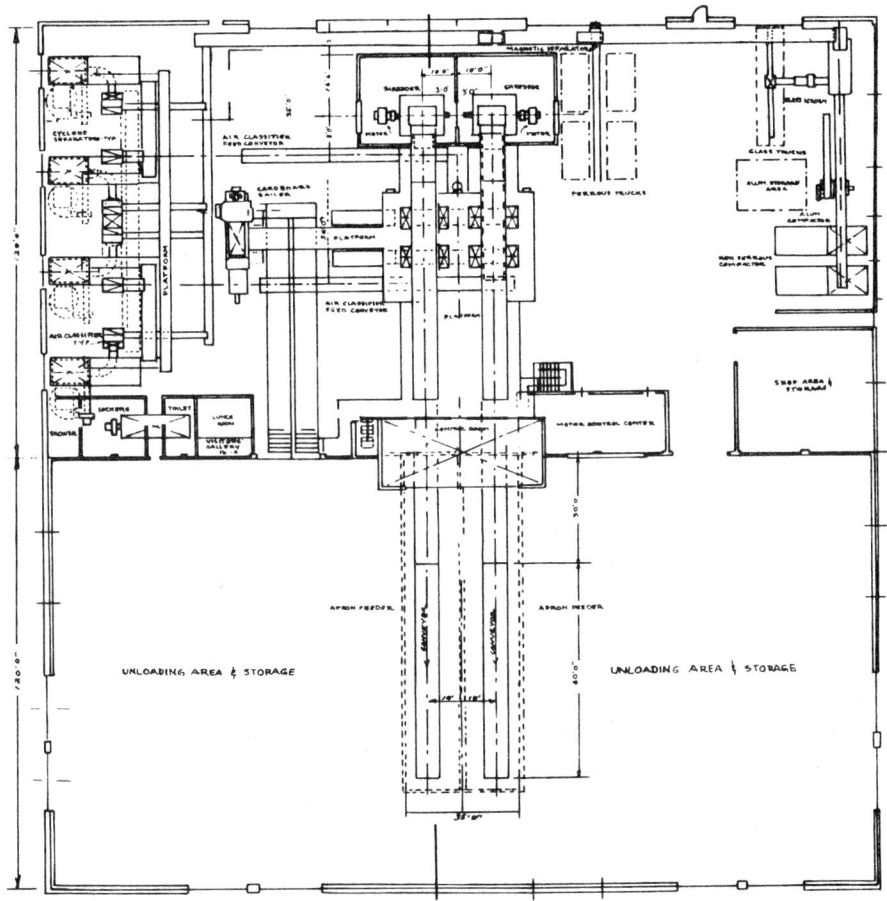

Figure 4-6
Tipping Floor
(Source: Americology)

Depressed Tipping Floor - This variation on the tipping floor concept utilizes an elevated area from where trucks dump onto a depressed flat floor; wheeled or tracked vehicles manipulate the refuse in the depressed area. Advantages include separation of the vehicle unloading and refuse handling functions, increased storage capacity, refuse pre-sorting, and the possibility of refuse compaction by the tracked vehicles.

Live Bottom - Shown in Figures 4-7, 4-8 and 4-9 this type of storage facility consists of a large pit with either apron conveyors or hydraulic ram conveyors at its bottom.

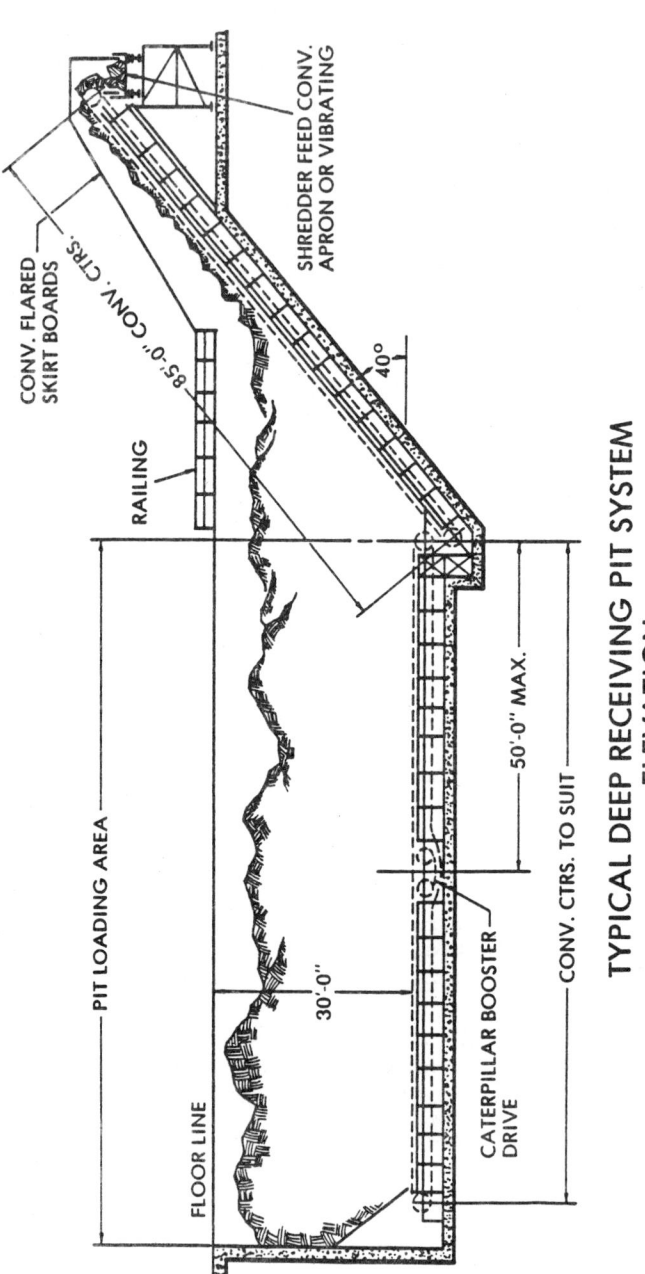

Figure 4-7
(Rexnord solid waste handling conveyors)

Unit Processes of Resource Recovery 61

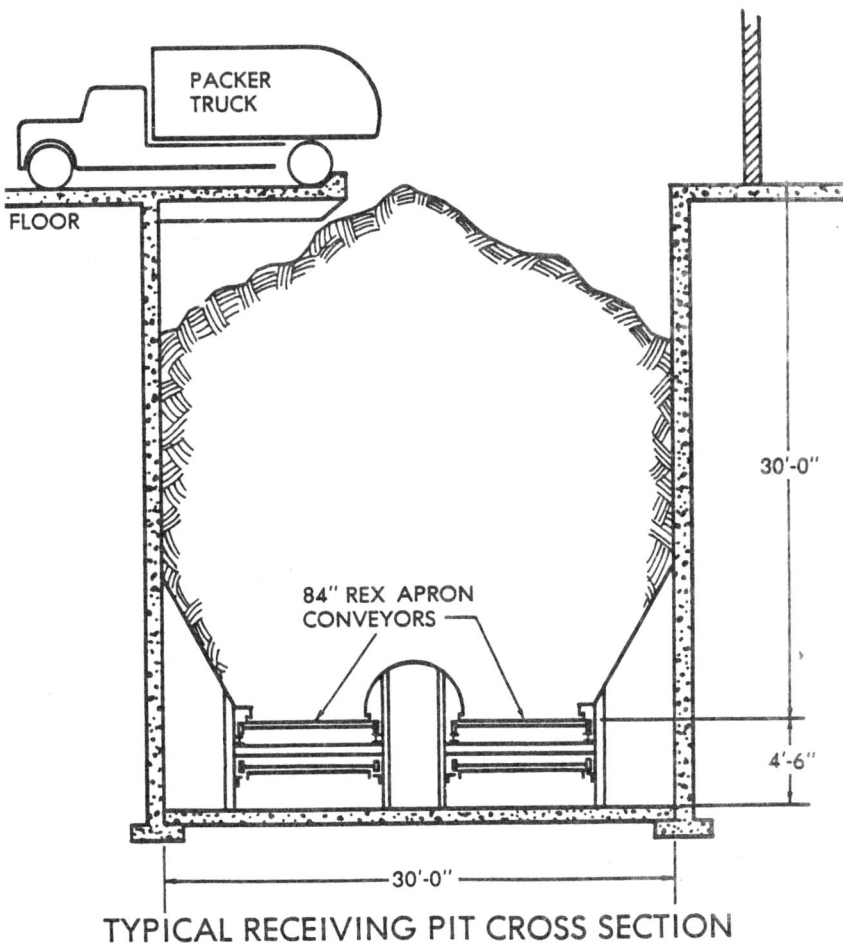

TYPICAL RECEIVING PIT CROSS SECTION

Figure 4-8
(Rexnord solid waste handling conveyors)

62 Energy and Resource Recovery from Waste

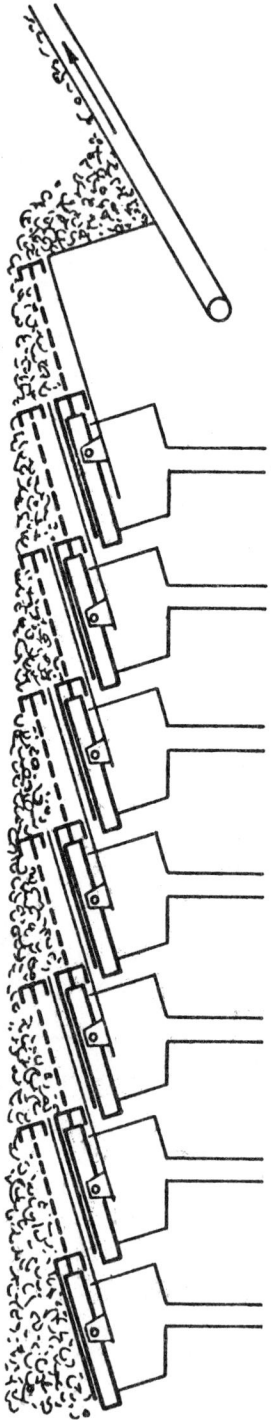

Figure 4-9
HYDRAULIC RAM CONVEYORS

Advantages include separation of unloading and refuse handling, automation of refuse handling, and large refuse storage capacity in reduced area. Disadvantages include high construction cost and difficulty in separating unprocessable and/or hazardous items. Another problem with this system is that refuse may not feed itself onto the conveyor but may bridge above the conveyor surface. There are few successful installations utilizing the live bottom system.

Example 4-3 illustrates the sizing of a refuse storage facility.

EXAMPLE 4-3

Consider a 1000 tpd resource recovery facility: size a pit and crane storage system.

From example 4-1, the facility must handle a peak of 88 truckloads/hour. If unloading takes 5 minutes:

$$\frac{\text{Truckloads/hr}}{\text{Bay}} = \frac{60}{5} = 12$$

$$\frac{88}{12} = 7.33 \text{ bays; use 10 bays}$$

10 bays x 15 ft/bay = 150 ft long

Assume width = 50 ft, density 350 lbs/cu yd., storage capacity 3 days.

1000 tons/day x 3 days = 3000 tons storage
3000 x 2000/350 = 17,143 cu yds.

$$\text{Depth} = \frac{17{,}143 \text{ cu. yds. x 27 cu.ft. per cu. yd.}}{150 \text{ ft. long x 50 ft. deep}} = 62 \text{ ft below tipping floor}$$

Trommel Screens

A trommel is a rotating perforated drum through which material is fed so that it can be classified by particle size. Trommels were first used in the mining industry to remove ore from soil approximately one hundred years ago. Today vibrating screens have replaced trommels in most industries. This is because, in general, when vibrating screens are utilized, the material being sorted is in greater contact with the vibrating screen surface and is more efficiently filtered out. Solid wastes, however, form a thick layer through which finer material cannot pass no matter how much the screen is agitated. Solid wastes are better suited to separation with trommels.[1,3,4] Figures 4-10 thru 4-12 illustrate typical

64   Energy and Resource Recovery from Waste

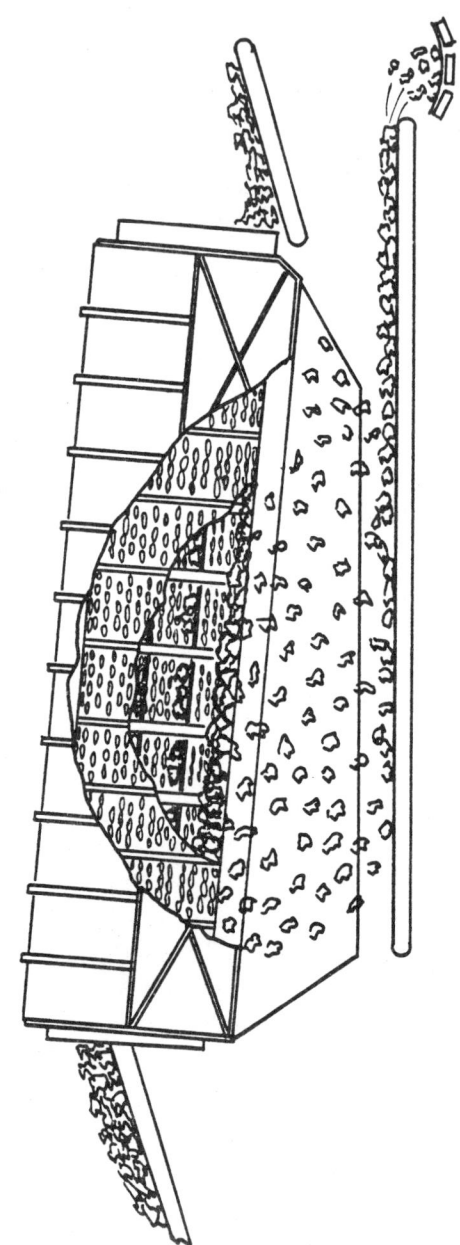

Figure 4-10
Trommel Screen
(NCRR: Reference 12)

Unit Processes of Resource Recovery 65

Figure 4-11

Trommel Screen
(NCRR: Reference 12)

66   Energy and Resource Recovery from Waste

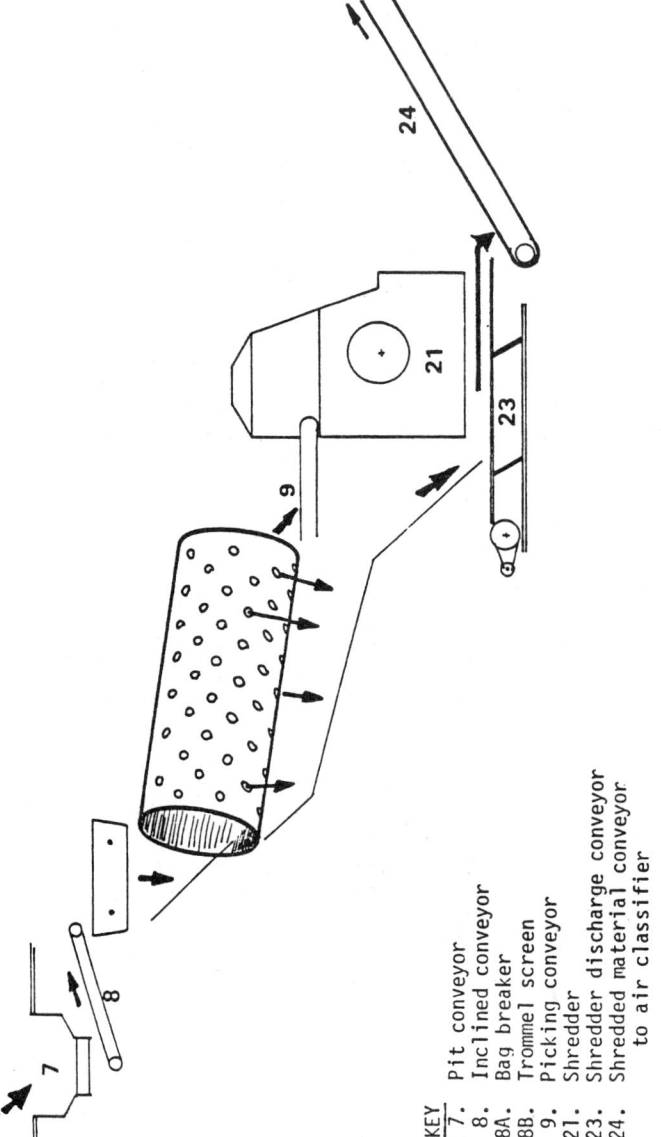

Figure 4-12
Trommel Screen
(NCRR: Reference 13)

KEY
7. Pit conveyor
8. Inclined conveyor
8A. Bag breaker
8B. Trommel screen
9. Picking conveyor
21. Shredder
23. Shredder discharge conveyor
24. Shredded material conveyor to air classifier

trommels and trommel installations. The inside face of the drum may have protruding knives or cutters which facilitate the breaking of plastic bags. The size of the perforations varies depending upon the type of material being classified. The largest trommel installed and operating today is at Recovery One in New Orleans. This unit is 45 feet long and the drum is 10 feet in diameter and has 4 3/4-in diameter holes.

In many installations, incoming wastes to a resource recovery facility were first fed to a hammermill or grinder, to break open bags of wastes and reduce the refuse in size. However, recent practice is to first feed the waste to a trommel, so that waste particles already smaller than shredder output are bypassed around the shredder.

The advantages of first feeding shredded wastes to a trommel include:

o   Reduced shredder capital costs since a smaller unit may be utilized.

o   Reduced maintenance on shredder parts since many abrasive materials bypass the shredder.

o   Higher BTU value in RDF since glass, metals and other non-combustibles are removed by the trommel.

o   Higher and more efficient recovery of aluminum cans and glass.

A trommel may also be utilized after shredding to concentrate recoverable material such as aluminum and glass. Trommels are used for screening of the heavy fraction after an air classification process for better recovery of aluminum and non-ferrous materials. Generally, a trommel used after shredding is enclosed to prevent dust from escaping.

The requirements of a trommel are generally set by the type of wastes being screened. The parameters of importance for optimum efficiency include:

o   rotational speed of trommel

o   slope of base frame

o   retention time in trommel

o   feed rate of wastes

If the rotational speed of the trommel is too slow, the wastes will not be sufficiently turned and the smaller particles will have less chance to reach the screen and filter out.

Also, slower speeds do not allow the wastes to reach the top of the trommel drum and fall down toward the screen to break up the waste into smaller pieces. One study conducted by NCRR concluded that a rotational speed of 11 to 12 rpm is optimum.[5] However, slower speeds will reduce dust problems from unenclosed drums. At too high a rotational speed, the wastes tend to centrifuge and not fall out. The slope of the drum can be varied to suit the particular wastes being screened. A steep slope will not allow the wastes sufficient time within the drum and a mild slope will not provide optimum throughput. Most units have variable slope frames so that the slope can be adjusted to achieve the best results.

The most important parameter for optimum screening efficiency is retention time in the drum. This should be set between 30 seconds and one minute and can be adjusted by varying the rotational speed, slope and throughput.

The optimum feed rate can vary for removing a specific item and testing is required for specific installations to optimize this parameter.

Shredding

Shredding is a term commonly used to define the size reduction operation at a solid waste facility. Size reduction of incoming solid wastes is desirable to ease the handling of the wastes and to increase the burning efficiency of the incinerator. When wastes are received at a facility they range in size from bulky oversized materials to small particles. Shredding reduces the volume of material being handled, promotes more efficient handling of the wastes for other processes and prepares the waste for air classification. Size reduction also increases the surface area of the wastes and promotes more efficient burning.

Size reduction is not a new technology. The mining and rock-crushing industries have used size reduction equipment to crush stone and ore for many years. However, for these applications, the new material is relatively uniform in size and is processed to a smaller size. Municipal solid waste is not uniform in size, therefore, the technology required is somewhat different. Generally only dry mechanical methods, (shredding, grinding or pulverizing), are used for solid waste size reduction. Only one system in current use, Black-Clawson's Hydrosposal process, utilizes wet shredding (hydropulping).

Unit Processes of Resource Recovery   69

Hammermills, both the horizontal shaft and vertical shaft type, are the most common equipment used for size reduction of municipal solid wastes. The horizontal mill has bottom grates and the rotating shaft is horizontal. Figures 4-13 and 4-14 show a typical horizontal mill. The grates on the bottom control the particle size of the output. Some installations also have grates at the top to control the particle size of the feed so that shutdowns and failures from oversized materials are avoided.

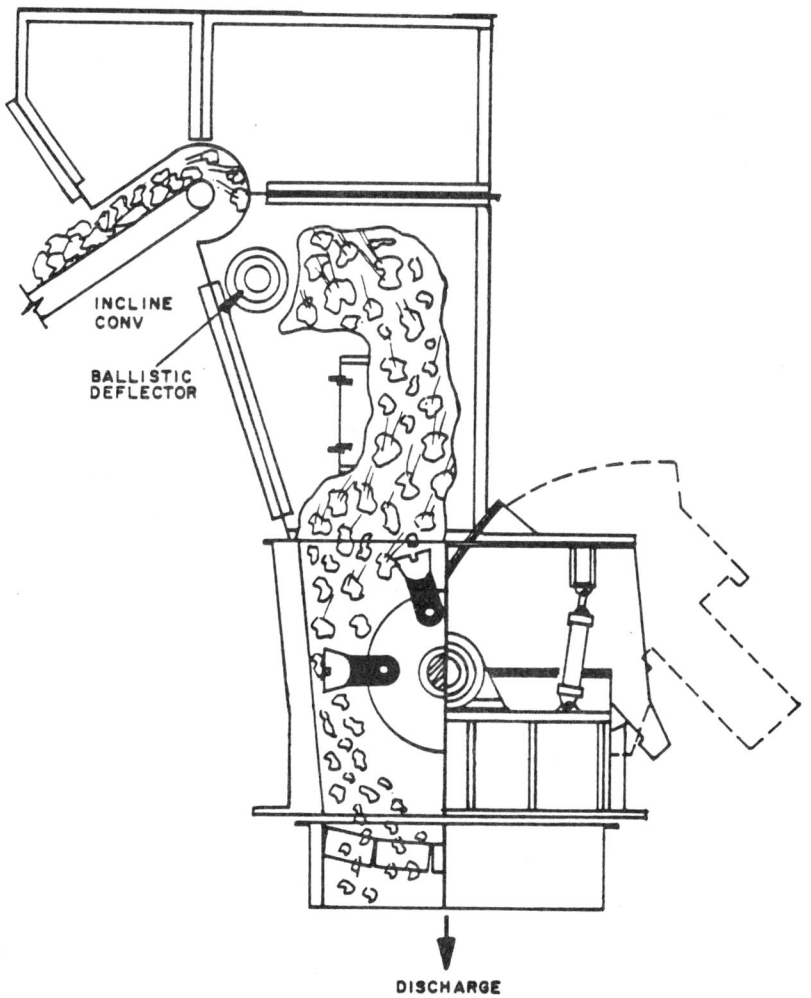

Figure 4-13
Gruendler Shredder
(NCRR: Reference 12)

Figure 4-14

Gruendler Shredder

(NCRR: Reference 12)

Unit Processes of Resource Recovery   71

Figures 4-15 and 4-16 present a typical vertical shaft hammermill. The input is fed at the top and the rotating shaft is vertical. Size reduction is controlled by the spacing between the side walls and hammer on the rotary shaft. The mill is tapered such that the bottom is smaller than the top.

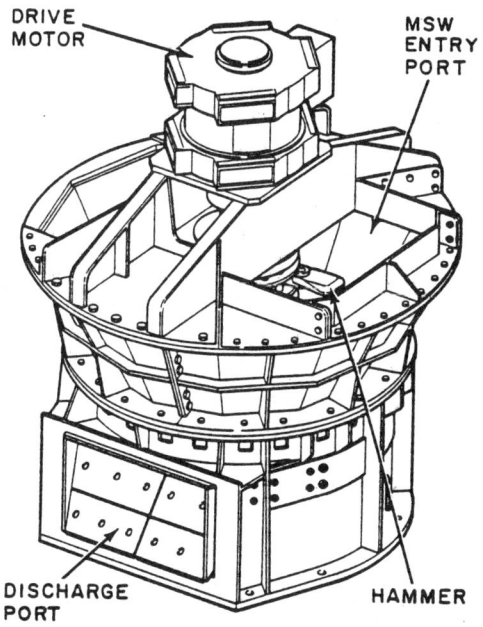

Figure 4-15

Heil Shredder

(NCRR: Reference 12)

Both types of hammermills operate at approximately the same power efficiency; however, there are differences between the vertical and horizontal mill.[6] Table 4-1 summarizes those differences.

72 Energy and Resource Recovery from Waste

Figure 4-16

Heil Shredder

(NCRR: Reference 12)

TABLE 4-1

COMPARISON OF HORIZONTAL AND
VERTICAL SHAFT HAMMERMILLS

Parameter	Horizontal Shaft	Vertical Shaft
Particle size control	Positive control; All material processed until small enough to pass grates	Not positive control; Particle size controlled by feed size;
Shredding of difficult items	Subject to damage from oversized and difficult material; therefore presorting and/or top screens desirable	Less subject to damage; Difficult items pass through machine: presorting not necessary
Wear rates	Hammer wear rates may be higher	Particle size gradually reduced so that wear on hammers is more evenly distributed
Motor Drive	Most manufacturers recommend direct drive	Most manufacturers recommend a gear or belt drive

There are two types of hammers available for either type of hammermill. The swing type hammer is commonly used for solid waste size reduction. The hammer is mounted on pins and is free to rotate which results in less damage to the shredder when difficult material is encountered. However, because the hammer rotates freely, the possibility exists that objects will become entangled and create an imbalance problem. Figure 4-17 [7] presents some common hammer shapes.

The selection of a shredder will depend upon the particular application and the type of input material. The size and power required for the machine will also depend upon the input material and the desired output particle size. Table 4-2 lists the minimum horsepower required for size reduction for municipal solid wastes.

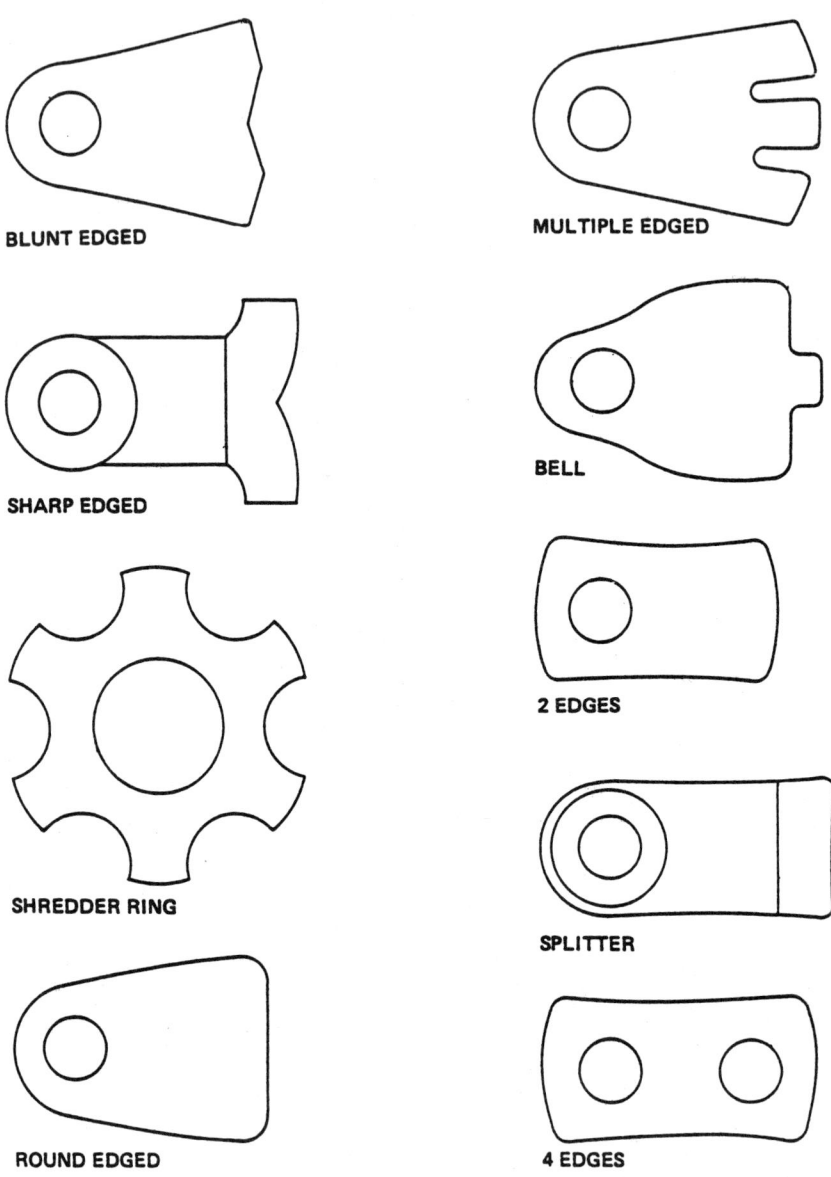

Figure 4-17: Types of hammers used on hammermills depicting their shapes. The weight may vary from 15 to 500 pounds each. Source: Funk, H.D. Henningston, Durham, and Richardson, Inc. Unpublished data, 1974.

TABLE 4-2 [6]

MINIMUM SHREDDER HORSEPOWER REQUIREMENTS

Solid Waste Category	Composition	Minimum Horsepower
Light	Sorted packer truck wastes, such as-paper, cardboard, bottles, cans, garbage, and lawn trimmings.	250
Medium	Normal packer truck wastes, such as-the above plus small crating, small appliances, small furniture, bicycles, tree trimmings, and occasional auto tires.	400
Bulky	Oversize and bulky wastes of the above plus items such as-stoves, refrigrators, washers, dryers, doors, large furniture, springs, mattresses, tree limbs, and truck tires.	800
Heavy	Large and dense materials, all of the above plus items such as-demolition rubble, automobiles, logs, and stumps.	2,000

The desired particle size of the output also effects the power requirements of the shredder. Figure 4-18 [6] is a plot of the particle size versus required horsepower. At resource recovery facilities, both a minimum and maximum particle size of the output is usually specified. When a large portion of the input contains oversized or bulky material and the required particle size of the output is small, two-stage shredding may be required. Manufacturer's data indicate that for those types of waste a particle size between two and four inches is economically feasible with one stage shredding.

Example 4-4 illustrates shredder selection.

The throughput of the shredder is also an important selection parameter. Generally, size reduction is not a 24-hour operation because of the time lag between collection of waste and delivery to the facility. For small plants (150-500 tons/days) an eight hour shift is generally recommended. At large plants, a 12-14 hour a day operating time would be more efficient. Allowance for down time for maintenance, adjustments and repairs should also be included.

A shredder selection nomograph is presented in Figure 4-19. [6]

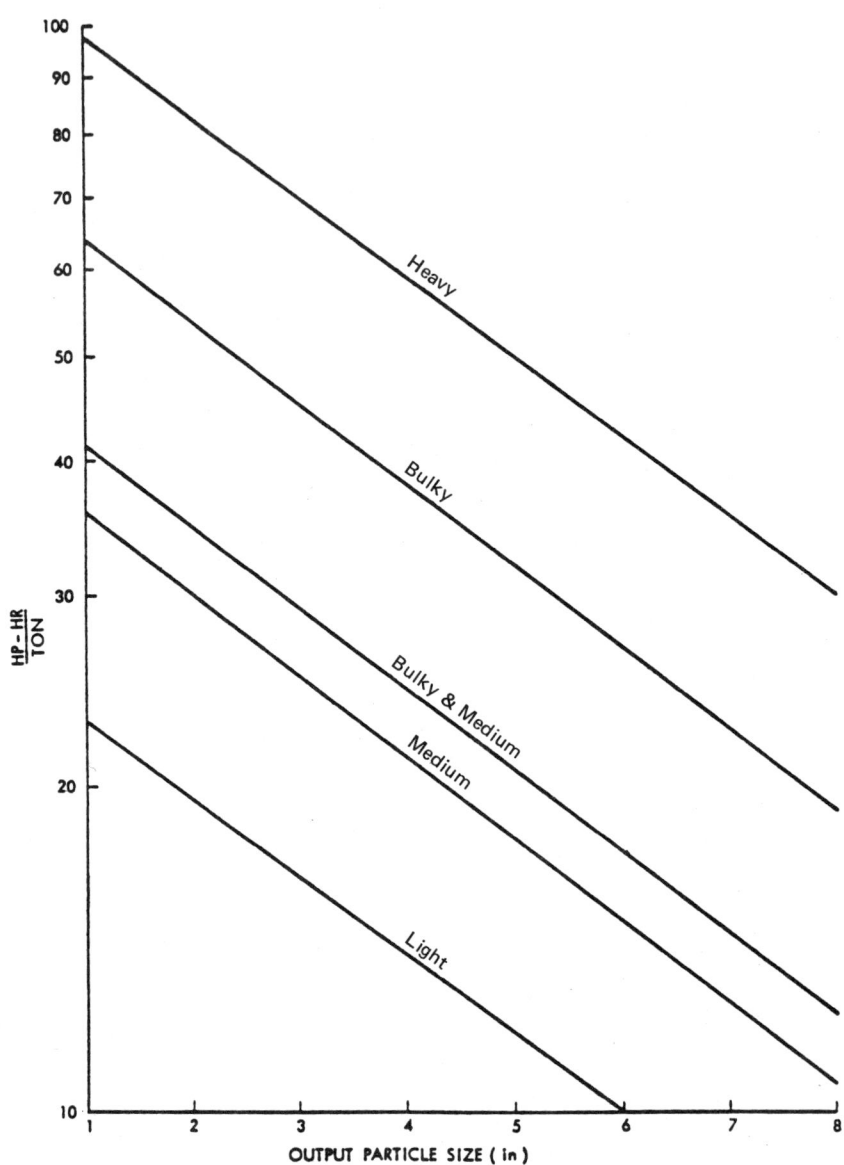

Figure 4-18: Solid Waste Size Reduction Power Requirement

EXAMPLE 4-4

Consider a shredder handling 50 tons per hour of combined bulky and medium solid waste, and producing a nominal output particle size of 3 inches. Calculate required horsepower.

Required Hp-Hr/ton for bulky and medium waste, 3-inch output size, Figure 4-18 is 30

$$\frac{30 \text{ Hp-Hr}}{\text{ton}} \times \frac{50 \text{ ton}}{\text{hr}} = 1500 \text{ Hp}$$

If nominal output particle size is increased to 6 inches what is required horsepower?

Required Hp-hr/ton from Figure 4-18 is 18

$$\frac{18 \text{ Hp-hr}}{\text{ton}} \times 50 = 900 \text{ Hp}$$

The design of the shredder should also include provisions for handling and storage of the wastes. The input is usually fed to the shredders by a conveyor and the output is brought to the next processing step by conveyor. The aspects of both the feed and discharge conveyors are discussed in another section. There are three different types of shredder installations:

- o At ground level
- o Below ground level
- o On a sloping site

For the shredder to operate effectively, sufficient clearance is required such that the feed to the conveyor has sufficient drop. When the shredder is installed at grade, an inclined conveyor which lifts the wastes up to the top of the shredder is required. This type of installation is the most common. By installing the shredder below grade, problems with inclined conveyors are avoided, however, access to the shredder for maintenance and repair is hindered. Where site conditions allow, the installation of a shredder at the side of a hill is preferred. Problems with inclined conveyors are avoided and access is not limited by below the ground installation. Figure 4-20[8] presents some typical shredder installations.

In addition to the feed and discharge systems, provisions for dust, explosion and fire control are necessary at shredder installations. Water spray systems have been used success-

78   Energy and Resource Recovery from Waste

fully to control dust and fires at several installations; this method has the disadvantage of increasing refuse moisture content which is counterproductive if incineration is intended.

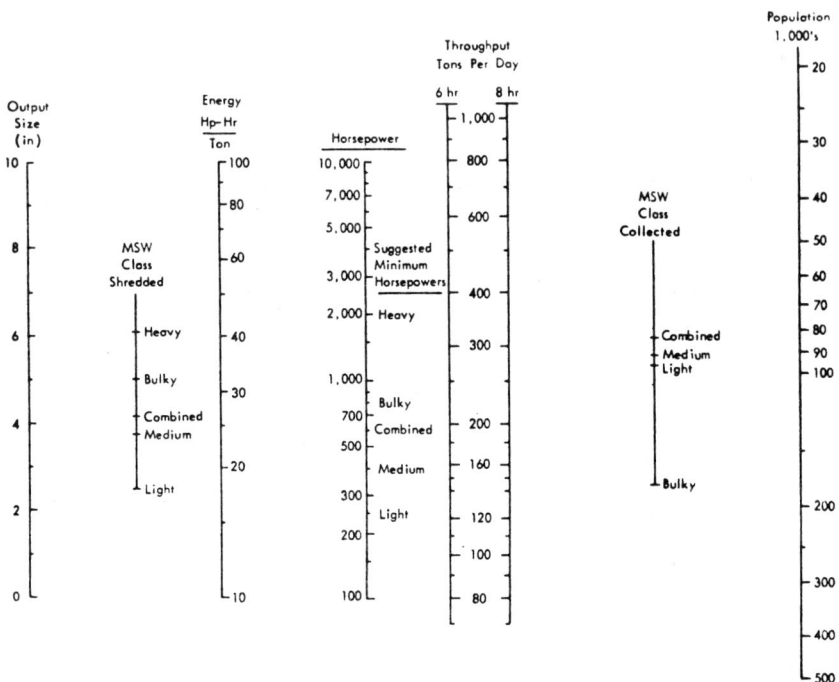

Use of nomograph:

A. Draw a line from POPULATION through MSW CLASS COLLECTED and intersect 6 hr or 8 hr THROUGHPUT.

B. Draw a line from OUTPUT SIZE through MSW CLASS SHREDDED and intersect ENERGY.

C. Draw a line between the intersection with ENERGY and the intersection with THROUGHPUT. The intersection with HORSEPOWER is the horsepower required to shred the indicated input to the indicated size range.

Figure 4-19
Shredder Selection Nomograph
(REFERENCE 6)

Unit Processes of Resource Recovery   79

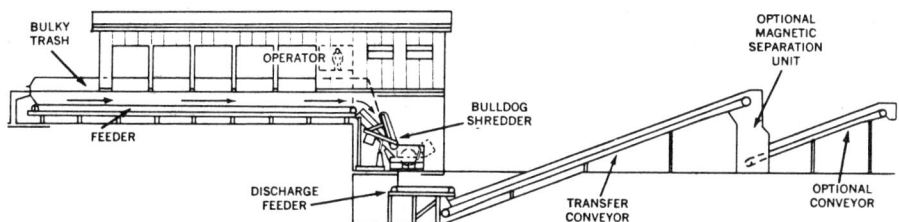

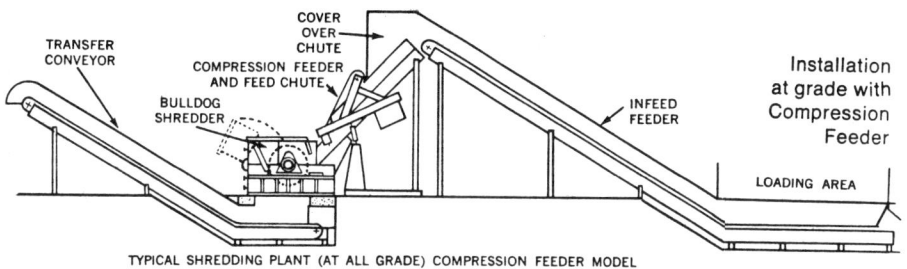

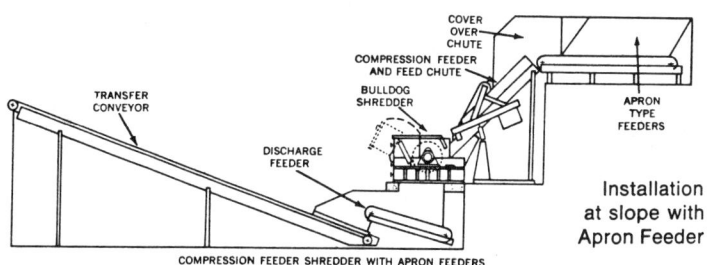

Figure 4-20
Typical Refuse Shredder Installations
(Hammermills, Inc.)

Another method to control dust and limit explosions is to continually exhaust the shredder housing with a pneumatic system. Such a system conveys the collected dust to a central point for storage and disposal.

Proprietary explosion suppression systems, such as the "Fenwal" system, also are available. Such systems operate by flooding the shredder housing with an inert gas in response to an extremely fast acting pressure sensor which can react to certain types of explosion before pressures reach destructive levels. In theory, these systems prevent explosions ("conflagrations") caused by substances, such as gasoline, which require air as a source of oxygen. Explosions caused by substances which include their own oxidant, such as TNT, cannot be prevented by the Fenwal system.

Air Classification

Air classification is used to separate components in the solid waste stream by weight/drag ratio. In practical terms this involves separating shredded solid waste into a "light" fraction consisting primarily of paper, plastic and other light organic material; and a "heavy" fraction consisting of heavier organic and inorganic material.

Air classifiers generally operate by allowing the waste stream to fall through a rising current of air. Light particles with large surface areas are lifted by the air current; heavy and/or dense particles fall.

Separation in an air classifier is a function of velocity and air flow. Low air flow and velocities will minimize carryover of undesirable inorganic particles, and maximize RDF quality. At the same time low air flow will lead to a loss of organics with the heavy fraction. High air flow rates have the opposite effect, maximizing RDF production rates but downgrading quality. A reasonable performance level for an air classifer is 85 percent recovery of light fraction with about 5 to 10 percent carryover of heavy organics and inorganics.

Carryover of glass fines is a problem in air classification since these particles increase the ash content of RDF and can cause slagging problems during combustion. A partial solution to this problem is the use of a trommel ahead of the shredder: this bypasses a large proportion of the glass around the shredder and minimizes the production of fines.

Figures 4-21 through 4-25 illustrate typical air classifiers.

Air classifier design must deal with two phenomena which tend to interfere with separation by weight/drag ratio:
- o the tendency for material to agglomerate; and
- o the tendency for heavy material to ride on top of high aspect ratio light material.

Air classifier designs include turbulence producing elements (zig-zag air path, impulse air feed, tumbling action, etc.) to counteract these tendencies by causing clumps of materials to break up.

The air classifier is but one component in the air classification system which includes a blower, a cyclone separator to remove entrained material from the air stream, and air pollution control equipment (wet scrubbers or bag house filters) to remove dust from the cyclone discharge.

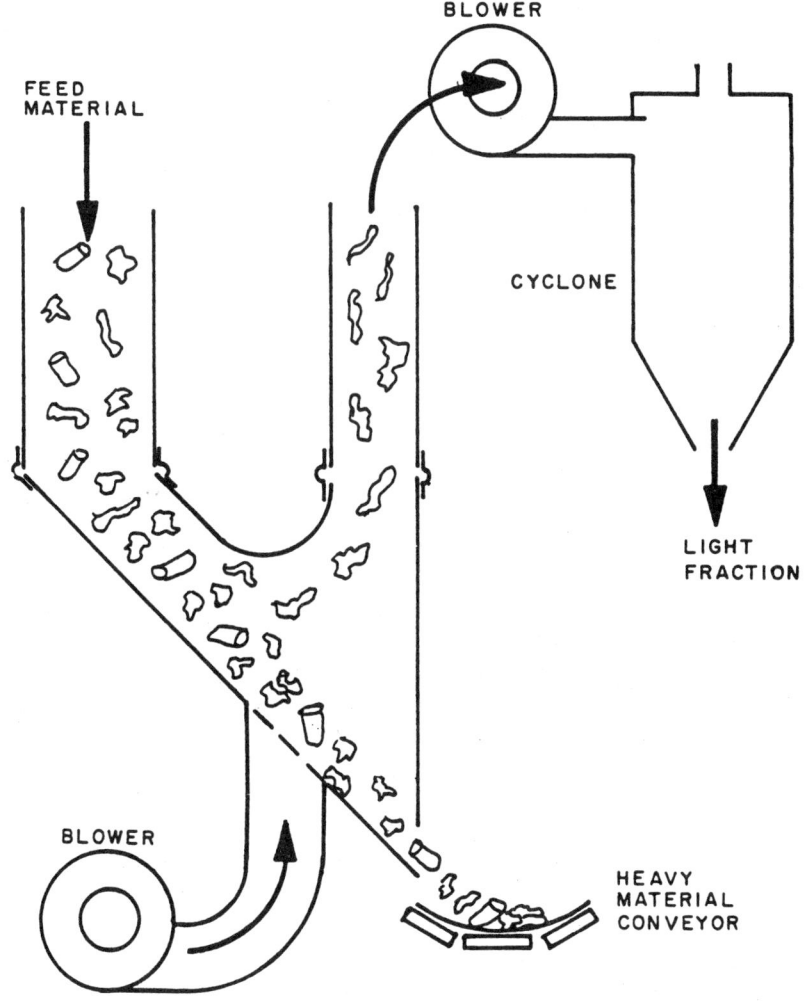

Figure 4-21

Air Classifier
(Triple/S)

(NCRR: REFERENCE 12)

82   Energy and Resource Recovery from Waste

Figure 4-22
Air Classifier (Triple/S)
(NCRR: REFERENCE 12)

Unit Processes of Resource Recovery   83

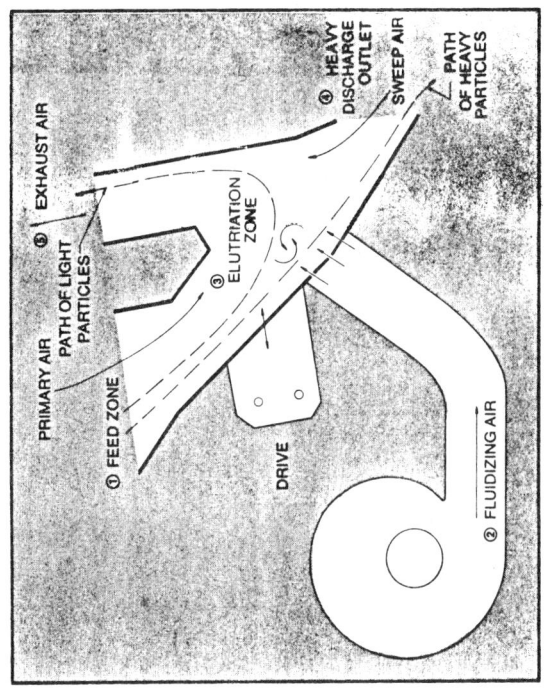
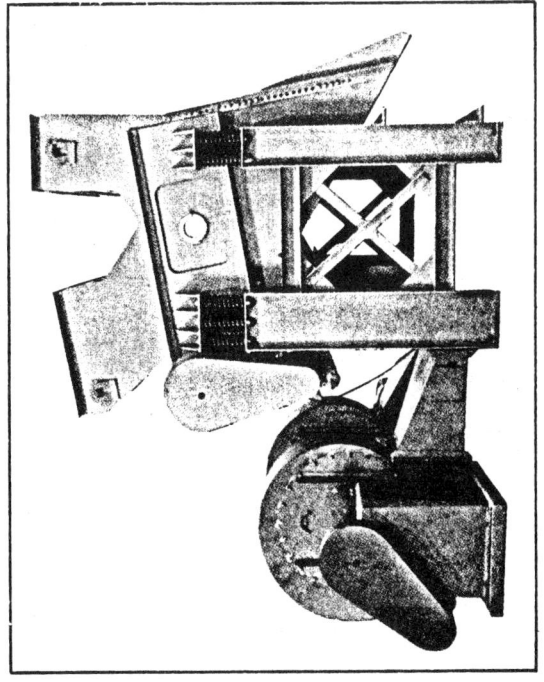

Figure 4-23
Vibrolutriator Air Classifiers
(Triple/S Dynamics)

84 Energy and Resource Recovery from Waste

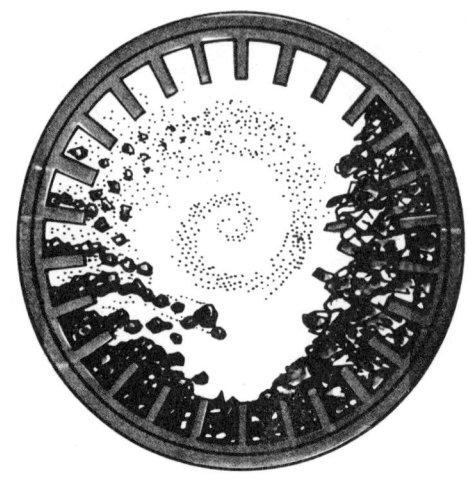

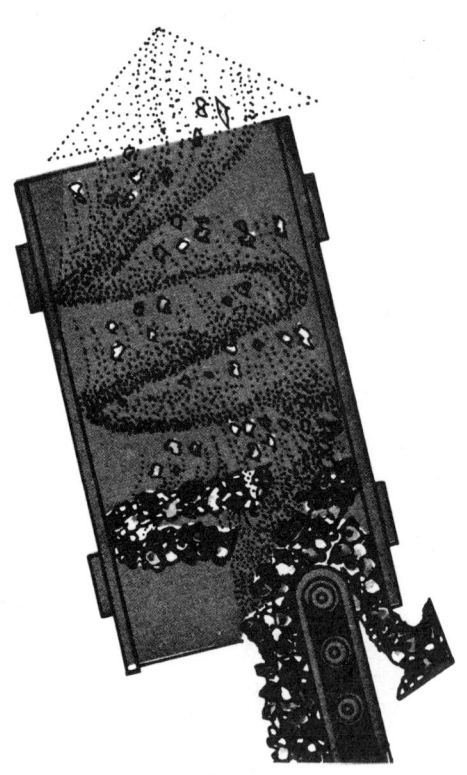

Figure 4-24
Rotary Drum Air Classifier
(Iowa Mfg. Company)

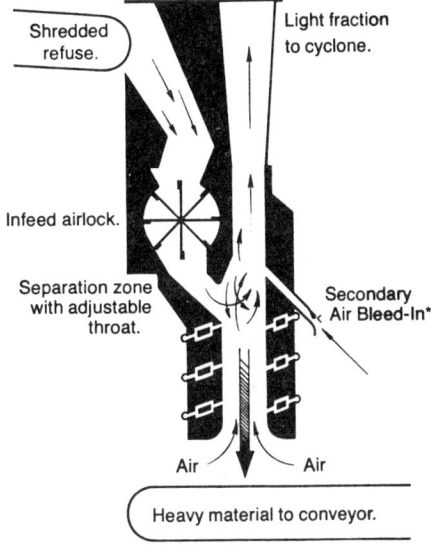

Figure 4-25
Air Classifier
(Rader)

ENERGY RECOVERY SYSTEMS

The predominant process for recovery of energy from waste in the United States is incineration.

Incineration is the destruction of waste materials by the controlled application of heat. There are a wide variety of incinerators that have been developed for the burning of the varied types and quantities of wastes found in today's society. Specialty incinerators have been developed for radioactive wastes, off-the-shelf units for hospital or apartment house waste, incineration equipment for burning ordinary household refuse, industrial incinerators for the destruction of liquid waste, and so on. Some of these incinerators have heat recovery equipment that generate steam, hot water or hot air, and some do nothing more than burn their designated waste streams. There are incinerators that are versatile in their application, able to burn anything from waste gas to garbage, tars to water-borne wastes, all within one unit. Following is a description of some of the various incinerators available, commonly employing energy recovery through waste destruction.

CENTRAL DISPOSAL INCINERATION SYSTEMS

Central waste disposal refers to the process of routing waste from many sources to a centrally located single facility for the disposal, usually by incineration, of the waste. Municipalities utilize central disposal for domestic waste and many industries will send their waste products to a central facility for incineration, particularly when an industry has many separate plants in relatively close proximity to each other. The economics of scale can often justify provision of heat recovery equipment, usually in the form of hot water or steam generation, from central disposal systems, whereas it is not economically justifiable from smaller facilities.

Europe has seen more activity in the central disposal of waste than the United States, and for a longer period of time. Three European incineration system, characterized by their grate designs, have been marketed throughout the United States. There are also a number of American systems, derivative from coal handling technology, which are used for central facility incineration.

EUROPEAN SYSTEMS - the three predominant incineration systems developed by European manufacturers are the reciprocating grate (Figure 4-26), roller (drum) grate (Figure 4-27), and reverse reciprocating grate systems (Figure 4-28). Examining the reciprocating grate, Figure 4-26, refuse dries out on the first grate section, burns on the second series of grates, and burns out to ash on the third grate. Adjacent grate elements are fixed and moving. The action of the moving grates drives the refuse, or ash, forward, to a stationary section. Material moves from the fixed grate to the adjacent moving grate by the action of new material entering the fixed grate zone. This successive movement results in a semi-continuous forward motion of material towards the end or bottom of the furnace while providing turbulence of the waste to encourage efficient burning.

A system of reverse reciprocating grates is shown in Figure 4-26. As the grates move forward, and then reverse, the waste is subject to continual agitation and forward motion, down towards an extractor section that aids in the discharge of ash from the furnace chamber.

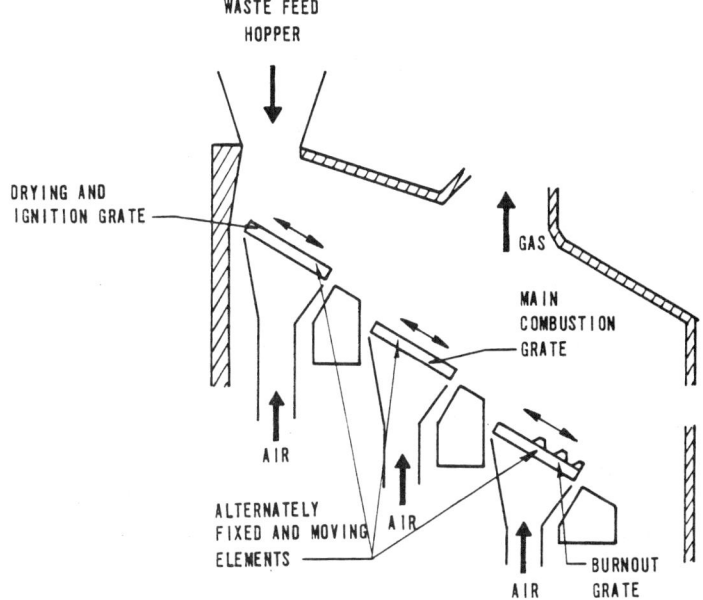

RECIPROCATING GRATE

Figure 4-26

Large hollow roller or drum grates are used in the system shown in Figure 4-27. The drums rotate slowly, gently agitating the waste and moving it along to subsequent drums. As in the other grate systems, air is provided beneath the grates for grate cooling and also as a supply of combustion air (underfire air) for the waste. The furnace walls are normally built of finned water tubes welded together, a "water wall" as shown in Figure 4-29, absorbing some of the furnace heat. The flue gases will exit the furnace chamber at approximately 1800 degrees Farenheit and will pass through a section of vertical boiler tubes across its flow path. An economizer is often provided after the main boiler tube sections to absorb additional heat from the gas stream to heat the boiler feedwater. An air preheater section may be provided to pre-heat combustion air entering the furnace by extracting heat from the exiting flue gas.

Typical European grate sizes, for standard grate "modules", are listed in Table 4-3.

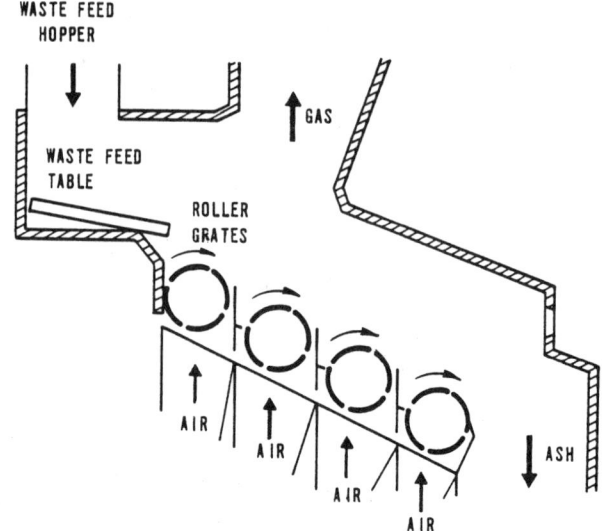

ROLLER (DRUM) GRATE

Figure 4-27

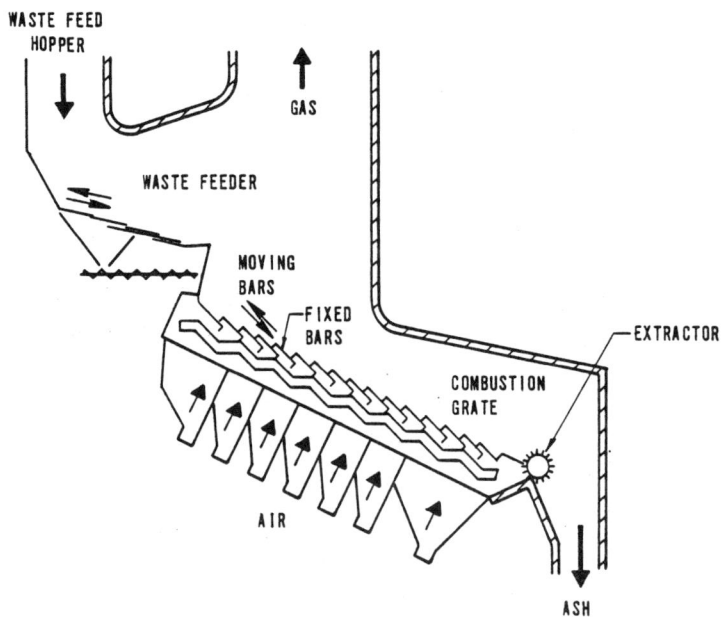

REVERSE RECIPROCATING GRATE

Figure 4-28

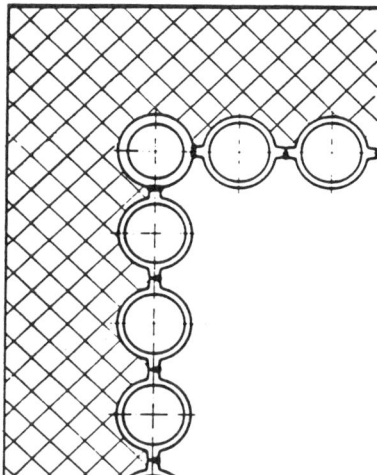

Figure 4-29
Sectional View Through Waterwall,
Showing Fin Welded Steel Tubing
(IBW-Martin)

TABLE 4-3

TYPICAL COMMERCIAL GRATES

Manufacturer	Module Capacity	Module Size	Loading
Wheelabrator (Von Roll)	750 TPD	15' W x 45' L	93 lb/hr ft^2
UOP (Martin)	150/170 TPD	6' W x 25' L	83/94 lb/hr ft^2
	1000 TPD	36' W x 25' L	93 lb/hr ft^2

Note: The external furnace size must include a minimum one foot refractory wall around the grate and three foot spacing between the end of the grate and the refractory wall for ash disposal.

AMERICAN SYSTEMS - the traveling grate, as shown in Figure 4-30, was the most common grate system manufactured in the United States for large waste incinerators until a few years ago. Two or three grate sections are normally used with drying taking place on the inclined grate section, burning on the second, and, if a third grate is provided, burnout to ash on that grate. Each grate is basically an endless belt of heavy interlocking cast sections, a variation of a conveyor system moving refuse across the bottom of a furnace chamber.

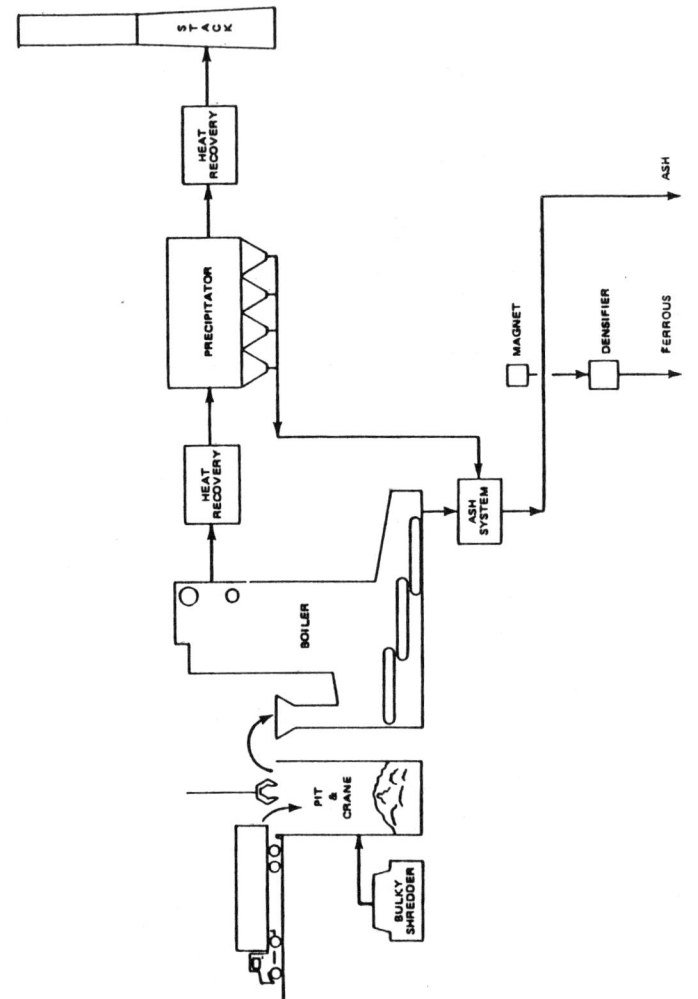

Figure 4-30
Typical Schematic for a
Mass Burning System
(Reference 10)

Another system gaining support in the United States is a suspension-fired system where waste is pneumatically charged into a furnace. Waste must be pre-processed, i.e., shredded, if this system is to have application. The European and travelling grate systems are all mass fired where no waste processing is necessary. The refuse hoppers, grates and ash disposal systems are all designed to handle bulk, heterogenous waste as received. Blowing refuse into a furnace so that most of the burning occurs within the furnace volume, and not on a grate, requires that the waste be shredded to relatively small mean size. Suspension fired systems normally have a single or duplex travelling grate system to catch those waste materials that have not been burned in suspension and to remove the ash generated from within the furnace.

The American, as well as the European systems, are normally built with water-wall construction and other boiler tube sections to maximize the generation of heat from the incinerator flue gas flow.

PYROLYSIS

There are a number of systems which utilize pyrolysis for waste destruction, a process related to incineration. Incineration requires that sufficient air (oxygen) be provided to adequately combust the waste to its basic combustion products, primarily carbon dioxide and water vapor. Pyrolysis is the destructive distillation of a solid material in the presence of heat and in the absence of air, or oxygen. It is an endothermic reaction, i.e., heat must be provided for the reaction to occur, whereas incineration will produce heat (exothermic).

A typical pyrolytic reaction is as follows, utilizing cellulose, the main constituent of paper:

$$C_6H_{10}O_5 \longrightarrow CH_4 + 2\ CO + 3\ H_2O + 3\ C$$

A gas is produced containing methane, CH_4, carbon monoxide, CO, and moisture. The carbon monoxide and methane are combustible, providing the off-gas with a positive heating value. The carbon residual, 3 C, a char, also has heating value.

Three major systems utilizing a variation of the

pyrolysis process have been marketed in the United States. A typical system is shown in Figure 4-31. Air is injected into the furnace to provide just enough burning to generate the heat required for the pyrolysis reaction to take place. (Stoichiometirc air, or oxygen, is that exact amount of air that is required for complete combustion of a combustible material. Pyrolysis ideally requires no oxygen but, as in this illustration, some air is added to the pyrolysis reactor. The air supplied is less than the stoichiometric requirement. In the burning process the ideal situation is never reached and air well in excess of the stoichiometric requirement must be provided to insure adequate burning, as much as 300% of the stoichiometric requirement when burning sludges or certain solid wastes.)

CONTROLLED AIR INCINERATION

The Controlled Air Incinerator is also referred to as a Modular Combustion Unit (MCU) or a Starved Air Combustion Unit. It consists of two combusion chambers, as shown in Figure 4-32. Waste is charged into the primary chamber and an amount of air less than the stoichiometric requirement is injected at a controlled rate. The thermal reaction in this chamber is a "starved-air" condition; not true combustion because less than the stoichiometric air requirement is introduced but not a pyrolysis reaction either because some air is injected. Sufficient air is added to the chamber to burn enough of the waste material to generate the heat required for maintenance of the process. The temperature within the primary chamber is usually maintained in the range of 1400 to 1600 degrees Farenheit.

Air is injected into the secondary chamber for combustion of the off-gas, which contains combustible organic material released from the waste within the primary chamber. Temperatures in the secondary combustion chamber can be as high as 2200 degrees Farenheit. With these high temperatures and with the large volume within the combustion chamber providing retention time for the hot gases, the exiting flow is relatively clean.

The small air flow injected into the primary chamber, where the waste is charged, results in less turbulence and therefore less carryover of particulate matter into the gas

Unit Processes of Resource Recovery 93

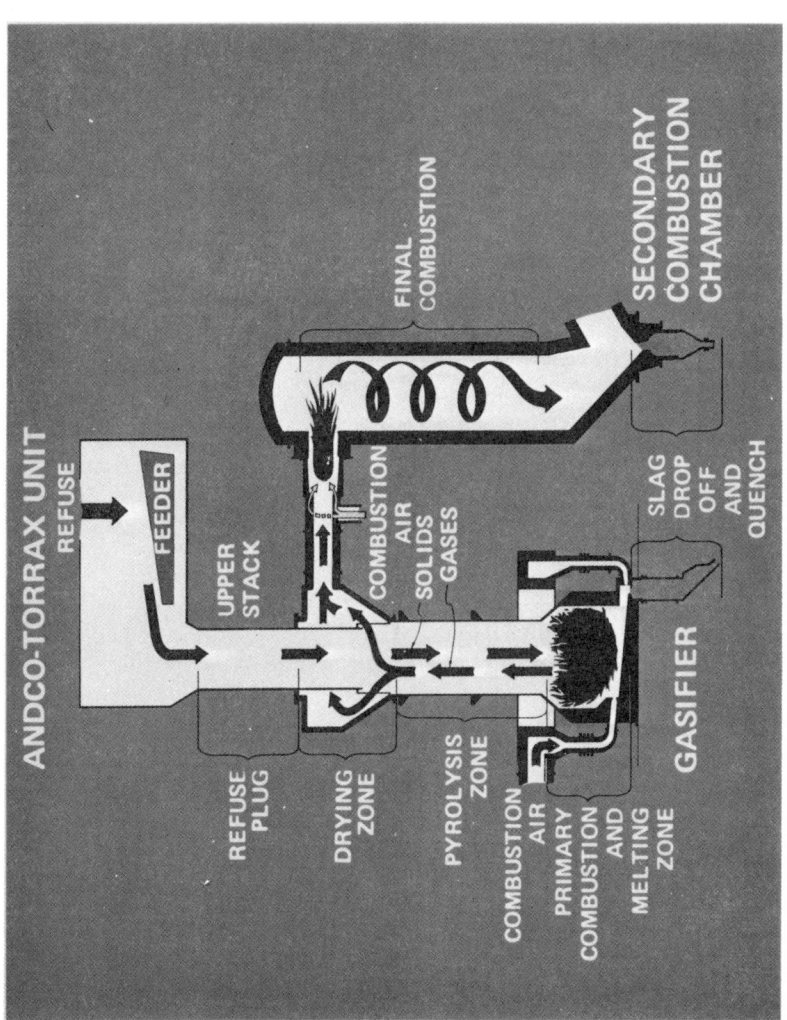

Figure 4-31
Andco-Torrax Combustion Unit
(Andco Inc.)

94 Energy and Resource Recovery from Waste

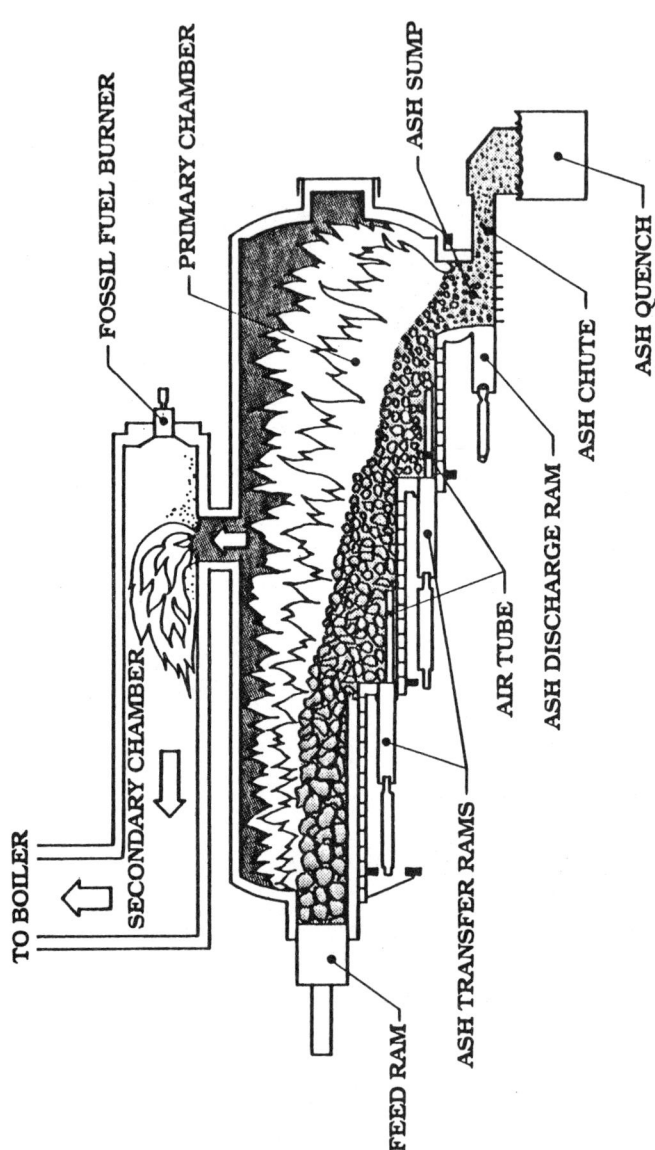

Figure 4-32
Typical Starved-Air Incinerator
(Consumat, Inc.)

stream than can be expected from incineration processes operating with air in excess of stoichiometric. Another feature of the starved air incinerator is the simplicity of its control. Air is introduced in response to the temperature in each chamber and this control is normally automated. The only other process control parameter is regulation of the quantity of waste fed to the unit.

ROTARY KILN

This incinerator is the most flexible and versatile of the incinerators in common usage today. It is widely used for the destruction of industrial wastes and has had application in the incineration of municipal solid waste and sludges. The rotary kiln is a horizontal refractory lined structure which rotates about its horizontal axis. Waste is fed from one side and by the action of the rotating chamber moves towards the end of the kiln. The kiln is sized and operated so that by the time the waste material reaches the end of the kiln it has burned out to an ash and the ash falls into a hopper for removal and disposal.

The kiln is mounted at an angle to the horizontal, or rake, and by varying the rake and the speed of rotation, the waste retention time within the unit can be varied. Rotation is normally variable within the range of 0.25 to 1.5 rpm. The length to diameter ratio of kilns used for waste disposal varies from 2:1 to 10:1 and the smallest size rotary kilns rarely are less than 36 inches in diameter.

Afterburner sections are normally provided with rotary kilns to insure complete burnout of the off-gases. Waste liquids and gases can be fired in the afterburner while solids and sludges are incinerated on the kiln hearth.

A variation of the rotary kiln is the O'Connor Combustor. This is a rotating horizontal device constructed of bare steel tubes. Water passes through the tubes, absorbing heat as the waste burns. Holes between the tubes allow injection of combustion or under fine air.

COMBUSTION CALCULATIONS

The combustible portion of a waste will burn to produce ash and a flue gas. The majority of ash in bottom ash, as opposed to ash particles airborne, exiting the furnace in the exhaust gas stream.

Products of combustion of the organic material in refuse will be carbon dioxide and water vapor (steam). The off gas will also contain oxygen and nitrogen with small quantities of sulfur oxides and trace amounts of other gases such as carbon monoxide, hydrogen chloride and nitrogen oxides.

Calculation of exhaust flow is based on the following parameters:

7.5 lbs of dry gas produced per 10,000 Btu released

0.51 lbs of moisture produced over 10,000 Btu released

These quantities represent the stoichiometric air supply, that theoretical quantity of air necessary for burning the combustible portion of the waste.

Example 4-4 illustrates a method of determining the products of combustion from incinerating MSW.

EXAMPLE 4-4

Determine the products of combustion from the burning of 250 tons per day MSW with 75% excess air.

Consider MSW as Class 1 wastes, then, from Table 2-7:

Moisture Content	25%
Incombustible Solids (Ash)	10%
Heating Value	6,500 Btu/lb as received

The actual input to the furnace, on an hourly basis will be determined:

Refuse 250 tons/day	=	20,833 lb/hr
less Moisture 25%	=	5,208 lb/hr
less Ash, 10%	=	2,083 lb/hr
Combustibles (moisture-ash-free)	=	13,452 lb/hr
Heating value	=	6,500 Btu/hr
Heat release (6,500 x 20,833 lb/hr)	=	135 MBH*

The flue gas generated by combustion at stoichiometric condition:

7.5 lb dry gas/10KB x 135 MBH = 101,250 lb/hr

.51 lb H_2O/10KB x 135 MBH = 6,885 lb/hr

The air required is calculated as follows:

Dry gas from combustion	101,250 lb/hr
Moisture from combustion	6,885 lb/hr
Sub-total	108,135 lb/hr

*MBH = million BTU per hour

less Combustible input	13,452 lb/hr
Stoichiometric air flow	94,683 lb/hr
plus 75% excess air	71,012
Total air required	165,695 lb/hr

The flow exiting the incinerator is composed of a dry gas component and a moisture component:

Dry gas from combustion	101,250 lb/hr
plus Excess air	71,012 lb/hr
Total dry gas flow	172,262 lb/hr
Moisture from combustion	6,885 lb/hr
plus Moisture from MSW	5,208 lb/hr
Total moisture flow	12,093 lb/hr

To calculate the temperature within the furnace, use Table 4-4.

For 172,262 lb/hr dry gas and 12,093 lb/hr moisture, assuming the dry gas has properties of air and taking the datum to be 60°F, and 3% radiation loss, the temperature at which the enthalpy of the gas flow is equal to the heat contained within the gas (135 MBH less 3% radiation equals 131 MBH in gas flow) is calculated as follows:

Assume a gas temperature of 2,300°F. The enthalpy of the gas flow, from Table 4-4, is:

Dry gas = 172,262 lb/hr x 596.45 Btu/lb = 103 MBH
H_2O = 12,093 lb/hr x 2252.60 Btu/lb = 27 MBH
Total = 130 MBH

Evaluating the flow at 2,400°F:

Dry gas = 172,262 lb/hr x 625.52 Btu/lb = 108 MBH
H_2O = 12,093 lb/hr x 2315.32 Btu/lb = 28 MBH
Total = 136 MBH

The temperature corresponding to a heat content of 131 MBH is between 2,300°F and 2,400°F. By interpolation, therefore:

$$\text{Gas temperatures} = 2{,}300 + \frac{(131 - 130)}{(136 - 130)} (2400 - 2300) = 2317°F$$

The flue gas volume (CFM) can be calculated using Table 4-5.

At 2300°F:

Dry gas = 69.5 ft^3/lb x 172,262 lb/hr ÷ 60 = 199,537 ft^3/min
H_2O = 111.9 ft^3/lb x 12,093 lb/hr ÷ 60 = 22,553 ft^3/min
Total = 222,090 ft^3/min

After passing through a waste heat boiler the gas would be reduced in temperature to, say, 400°F. The flow at this temperature is:

Dry gas = 21.7 ft^3/lb x 172,262 lb/hr ÷ 60 = 62,301 ft^3/min
H_2O = 34.9 ft^3/lb x 12,093 lb/hr ÷ 60 = 7,034 ft^3/min
Total = 69,335 ft^3/min

Flues are sized for no more than 3,000 feet per minute of flow. At 69,335 ft^3/min, the required flow area is:

$$\frac{69,335 \text{ ft}^3/\text{min}}{3,000 \text{ ft/min}} = 23 \text{ ft}^2$$

This area is slightly less than that of a square flue with an internal dimension of 5 feet.

Standard air (at atmospheric pressure and 70°F) weighs .075 lb/ft^3. The air duct size for the 166,240 #/hr air flow required, allowing the air velocity to reach 4,000 feet/min, is:

$$\frac{165,695 \text{ \#}}{.075 \text{ \#/ft}^3} \times \frac{\text{hr}}{60 \text{ min.}} \times \frac{1}{4000 \text{ ft/min}} = 9.2 \text{ ft}^2$$

or a square duct 37 inches on a side.

TABLE 4-4
ENTHALPY, AIR AND MOISTURE

Relative to 60°F			Relative to 80°F	
H Air BTU/#	H H$_2$O BTU/#	Temp. °F	H Air BTU/#	H H$_2$O BTU/#
21.61	1091.92	150	16.82	1071.91
33.65	1116.62	200	28.86	1096.61
45.71	1140.72	250	40.92	1120.71
57.81	1164.52	300	53.02	1144.51
69.98	1188.22	350	65.19	1168.21
82.19	1211.82	400	77.40	1191.81
94.45	1235.47	450	89.66	1215.46
106.70	1259.27	500	102.00	1239.21
119.21	1283.07	550	114.42	1263.06
131.69	1307.12	600	126.90	1287.11
144.25	1331.27	650	139.46	1311.26
156.87	1355.72	700	152.08	1335.71
169.59	1380.27	750	164.80	1360.26
187.38	1405.02	800	177.59	1385.01
195.26	1430.02	850	190.47	1410.01
208.21	1455.32	900	203.42	1435.31
221.25	1480.72	950	216.46	1460.71
234.36	1506.42	1000	229.57	1486.41
247.55	1532.40	1050	242.76	1512.40
260.81	1558.32	1100	256.02	1538.31
274.15	1584.80	1150	264.36	1564.80
287.55	1611.22	1200	282.76	1591.21
301.02	1638.26	1250	296.23	1618.20
314.56	1665.12	1300	309.77	1645.11
328.17	1692.15	1350	323.38	1672.15
341.85	1719.82	1400	337.06	1699.81
355.58	1747.70	1450	350.82	1727.70
369.37	1775.52	1500	364.58	1755.51
397.17	1832.12	1600	392.33	1812.11
425.08	1890.11	1700	420.29	1870.10
453.24	1948.02	1800	448.45	1928.01
481.57	2007.17	1900	476.78	1987.70
510.07	2067.42	2000	505.28	2047.41
538.72	2128.70	2100	533.93	2108.70
567.52	2189.92	2200	562.73	2169.91
596.45	2252.60	2300	591.66	2232.60
625.52	2315.32	2400	620.73	2295.31
654.70	2377.80	2500	649.91	2357.80
684.01	2443.30	2600	679.22	2423.30
713.42	2511.88	2700	708.63	2491.80

TABLE 4-5
SPECIFIC VOLUME

T °F	AIR FT³/#	H₂O FT³/#	T °F	AIR FT³/#	H₂O FT³/#
70	13.3	21.5	2000	61.9	99.7
100	14.1	22.7	2100	64.5	103.8
200	16.6	26.8	2200	67.0	107.8
300	19.1	30.8	2300	69.5	111.9
400	21.7	34.9	2400	72.0	116.0
500	24.2	38.9	2500	74.5	120.0
600	26.7	43.0	2600	77.0	124.1
700	29.2	47.0	2700	79.6	128.1
800	31.7	51.1	2800	82.1	132.2
900	34.2	55.1	2900	84.6	136.2
1000	36.8	59.2	3000	87.1	140.3
1100	39.3	63.3	3100	89.6	144.3
1200	41.3	67.3	3200	92.2	148.4
1300	44.3	71.4	3300	94.7	152.5
1400	46.8	75.4	3400	97.2	156.5
1500	49.4	79.5	3500	99.7	160.6
1600	51.9	83.5	3600	102.2	164.6
1700	54.4	87.6	3700	104.7	168.7
1800	56.9	91.6	3800	107.3	172.7
1900	59.4	95.7	3900	109.8	176.8

MATERIALS RECOVERY UNIT PROCESSES

Materials recovery incorporates the methods and procedures used to reclaim saleable materials from solid waste for return to the economy. There are two major methods of recovery - recovery from mixed waste at a resource recovery facility by handpicking and/or mechanical separation and source separation. This section deals with recovery at a resource recovery facility, source separation is discussed in another chapter.

In general, the mechanical methods used to segregate solid waste into economically valuable components are based upon techniques used in both the mining and the pulp and paper industries. These methods are aimed at recovering the purest product achievable in order to obtain the maximum dollar value for the recovered material.

Materials recovery to date has been concentrated in four areas: the recovery of paper (or fiber), ferrous metals, glass and aluminum.

The unit processes used to recover the marketable fraction of the solid waste stream will be described in this section. Process steps include:
- o Handpicking
- o Screening

- Shredding
- Ferrous material recovery.
- Glass recovery.
- Nonferrous material recovery.

Handpicking

Handpicking is the process of manually removing large, bulky items such as bundled paper or refrigerators which have value if recovered or which might cause damage to the shredders or other equipment. This initial step in separating the solid waste is performed by an operator positioned beside the conveyor system which is transporting the waste from the tipping floor or storage area towards the trommel or shredders. Any newspapers or corrugated boxes which have remained relatively intact and uncontaminated can be baled and sold for reuse. The quantity and quality of the paper will determine the economic feasibility of this process. Other practical considerations include:

- Positioning the operator a safe distance from the shredding equipment
- Provisions for an overhead crane, etc., for the removal of heavy objects
- Maintaining a maximum 36 feet per minute conveyor speed [12]
- Accessibility to depositories for the removed items.

Screening

Screening is a simple unit process that may be employed at any of the several stages in resource recovery processing. Screens are used to separate solid waste particles according to size. The screen opening may be varied as the purpose dictates, and the size should be designed for each specific use. Different methods of screening exist, including rotary screening with trommels and single-or multi-level vibrating screens.

<u>Trommels</u> - Trommels, discussed earlier in this chapter, are rotating cylindrical screens. The material to be processed is introduced into the inside of the cylinder, the rotation of which causes the material to tumble with particles smaller than the perforations of the trommel falling through. The tumbling action of the trommel makes it relatively free

from blinding (clogging) compared to a flat screen. However, trommels have higher initial purchase prices, higher replacement costs for screen surface and require more room for installation.

<u>Vibrating Screens</u> - Flat vibrating screens, both single and multi-level, can also be used to sort shredded waste by size. Such a screen is illustrated in Figure 4-33.

Shredders

In addition to their role in the primary size reduction of incoming solid waste, shredders are also used in materials processing to effect further size reduction of solid waste components. Examples include the use of hammermills to shred ferrous metals, rod mills to produce a glass cullet of uniform size and consistency, and flail mills, sometimes used to avoid "balling" cans.

Ferrous Material Recovery

The recovery of ferrous metals is accomplished using magnetic separators which attract the ferrous metals, but not the other components of the waste. There are two types of magnetic separators in use. One type (See Figs. 4-34 and 4-35) consists of a large revolving drum with magnets mounted inside. Each magnetic drum separator is positioned over the head pulley of the material conveyor such that ferrous metals are lifted onto a transporting conveyor and conveyed to a discharge point. This type of unit is manufactured by Stearns Magnetics, Inc.

The magnetic drum will not pick up any small magnetic particles lying underneath other materials in the waste stream, nor those pieces which are only partly ferrous. This material can be removed later by a scavenger magnet. The magnetic drum has been used in both the mineral processing industry and the scrap processing industry for many years.

The second major type of magnetic separator is the magnetic belt separator, shown in Figure 4-36. A magnetic belt separator has been developed specifically for materials recovery from solid waste by Dings, Inc. The Dings arrangement, which utilizes multiple magnets under the belt, is capable of producing a relatively clean ferrous product.

Figure 4-33
4 x 2 inch vibrating screen
(NCRR: Reference 12)

Unit Processes of Resource Recovery 103

Figure 4-34
Magnetic Drum Separator
(NCRR: Reference 12)

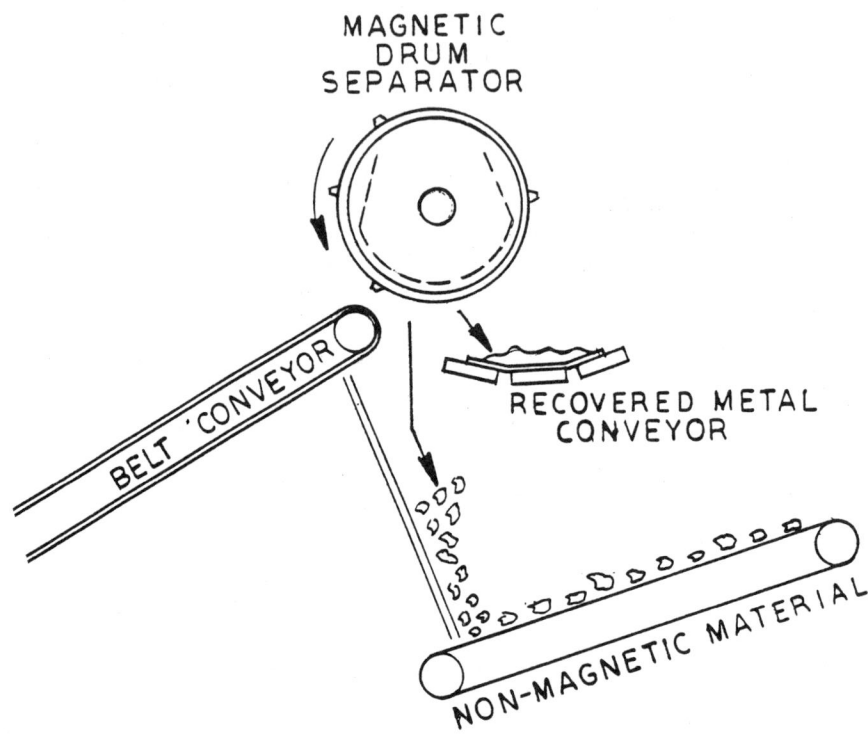

Figure 4-35

Magnetic Drum Separator

(NCCR: Reference 12)

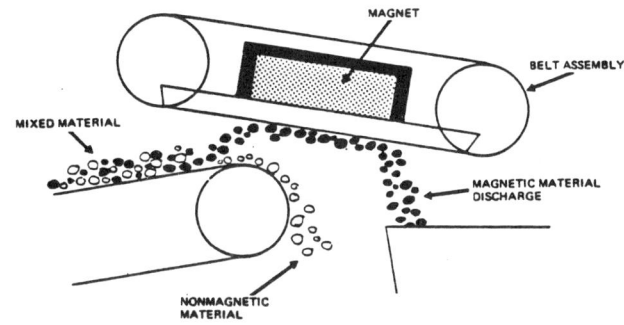

Overhead Belt Separator

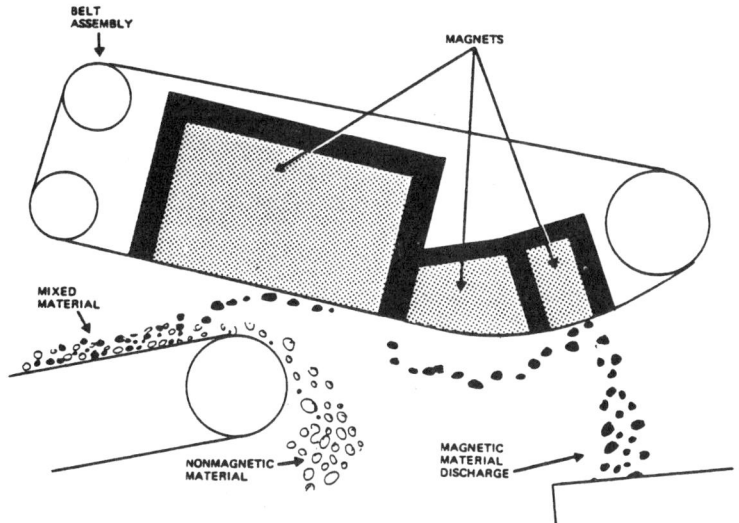

Dings Co. Belt Magnet Designed for Solid Waste Systems

Figure 4-36
Overhead Belt Separators
(NCCR: Reference 12)

One problem associated with belt separators is accelerated belt wear. Special attention to the belt specifications will minimize this situation. Magnetic separation efficiency ranges between 70 and 99 percent depending upon the process used.

Other suppliers of magnetic evaluation equipment for ferrous scrap are include Eriez Magnetics and Indiana General.

The ferrous metal product can be further cleaned of organics and divided into light and heavy fractions using a type of air classifier called an air knife separator.

Glass Recovery

Two methods for recovery of glass from MSW in a marketable form have been tested on a limited basis. These are optical sorting and froth flotation. Optical sorting is designed to produce glass fractions according to color (flint, green, and amber) while froth flotation produces a mixed-color glass fraction. Optical sorters have been used for a number of years in the food processing industry for sorting nuts and grains. Froth flotation has been used for more than 60 years by the mineral processing industry for beneficiation of ore. Figure 4-37 illustrates both systems. Either method would be installed in a resource recovery system following removal of metals and organics. The stream of material for processing at that point normally consists of glass, stones, and refractory type materials.

Optical Sorting - This process, shown in Figures 4-37 through 4-40, is restricted to sorting glass particles from 3/16 to 5/8-in. in size. Particles of this size can be individually inspected according to the amount of light transmitted to a photocell. Once categorized, the particles are separated pneumatically. In testing performed thus far with optical sorting, separation of the glass fractions by color has been successful. However, opaque particles such as stones have not been effectively rejected by this system, resulting in an inferior product for glass container manufacturing. An additional problem with optical sorting is the low recovery rate that may result from some of the glass fraction breaking into unacceptably small (less than 3/16-in.) pieces.

Froth Flotation - This process, shown in Figures 4-41 and 4-42 is appropriate for small-sized glass fines. It is based on the physical reaction of different surfaces to water. In practice, a small amount of air is blown through a mixture of glass and nonglass particles submerged in water; the glass, which is hydrophobic, clings to the air bubbles and floats, while other particles sink. The addition of inexpensive chemical reagents, such as fatty amines, increases the efficiency of the separation. Since the froth flotation process yields color-mixed glass fines, the product is limited to nonclear (green or amber) glass manufacturing, representing only 30 to 40 percent of glass container production.

Unit Processes of Resource Recovery 107

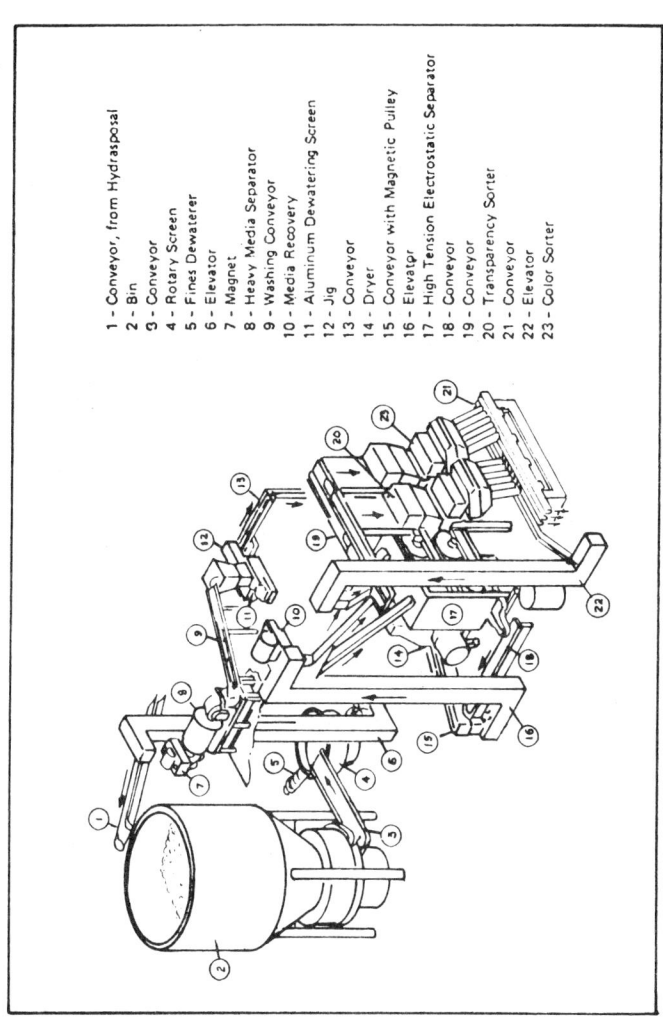

Franklin, Ohio Color Sorting

Figure 4-37

GLASS RECOVERY SUBSYSTEMS

1 - Conveyor, from Hydrasposal
2 - Bin
3 - Conveyor
4 - Rotary Screen
5 - Fines Dewaterer
6 - Elevator
7 - Magnet
8 - Heavy Media Separator
9 - Washing Conveyor
10 - Media Recovery
11 - Aluminum Dewatering Screen
12 - Jig
13 - Conveyor
14 - Dryer
15 - Conveyor with Magnetic Pulley
16 - Elevator
17 - High Tension Electrostatic Separator
18 - Conveyor
19 - Conveyor
20 - Transparency Sorter
21 - Conveyor
22 - Elevator
23 - Color Sorter

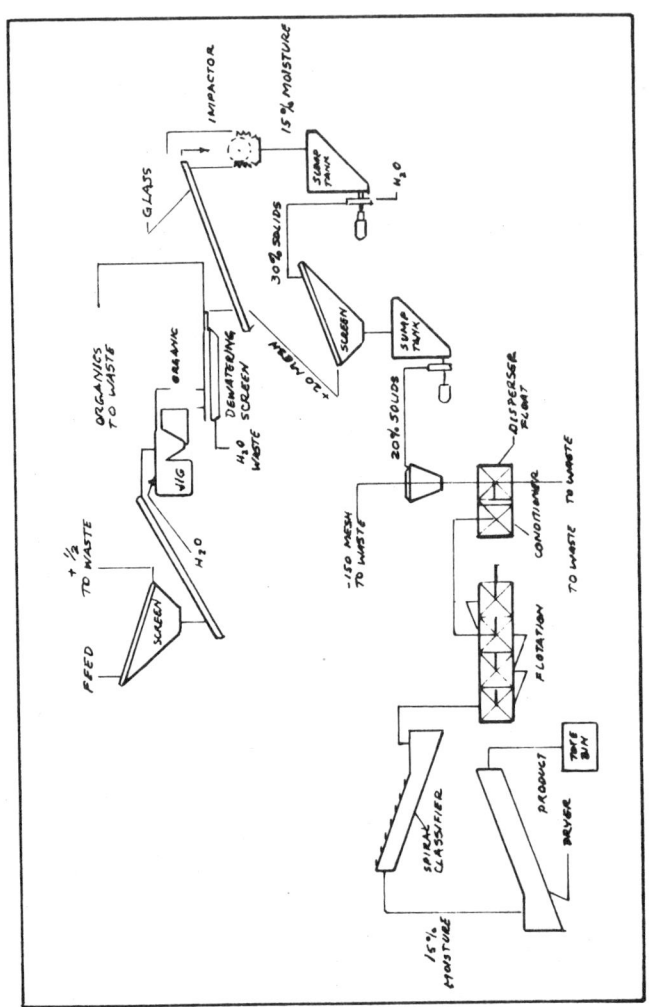

Figure 4-37 (continued)

Source: "Status Report on Glass Recovery," The Glass Industry, by H.W. Gershman, October, 1976.

Unit Processes of Resource Recovery   109

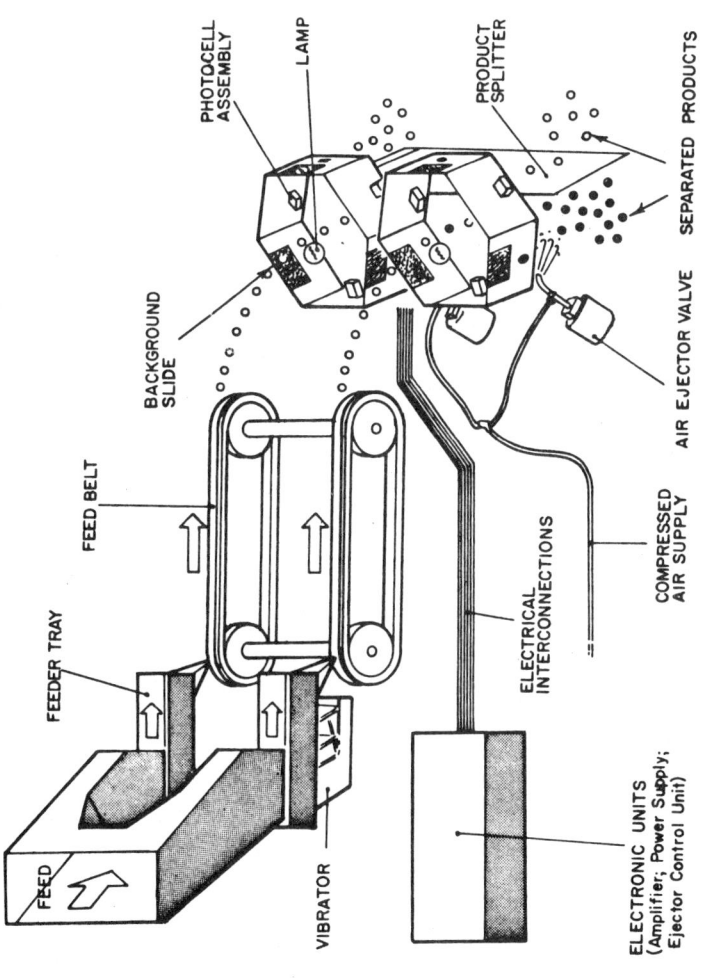

Figure 4-38
Glass Color Sorter;
Sortex Company of North America
(Reference 13)

Figure 4-39
Optical sorter separating
clear glass from colored glass;
Sortex Company of North America
(Reference 13)

Unit Processes of Resource Recovery 111

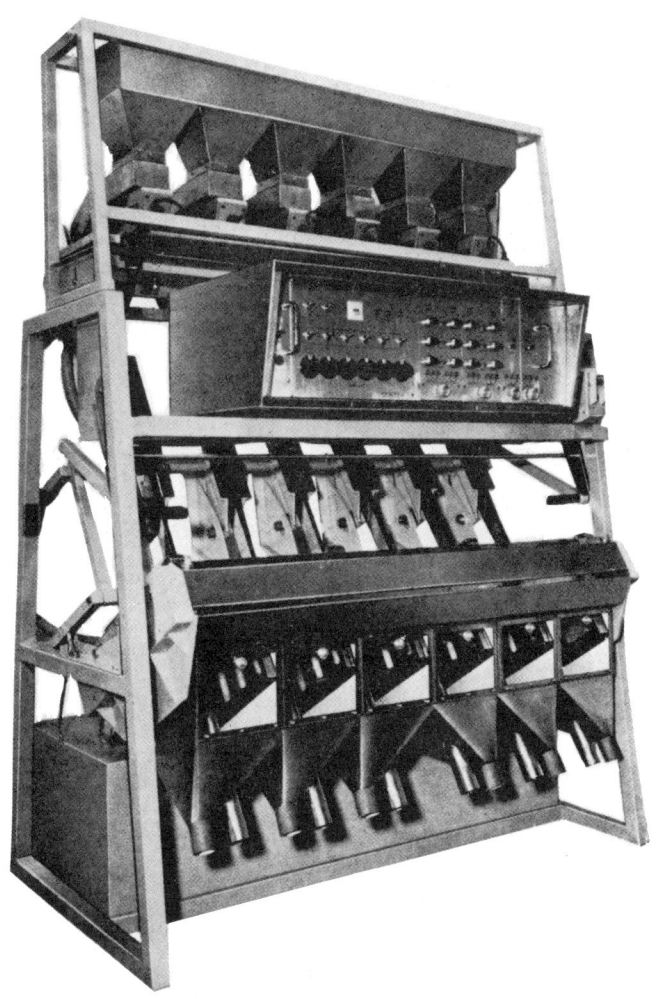

Figure 4-40
Sortex 962M Optical Separator;
Sortex Company of North America
(Reference 13)

Figure 4-41
Flotation Cell Unit
(NCRR: Reference 12)

Unit Processes of Resource Recovery 113

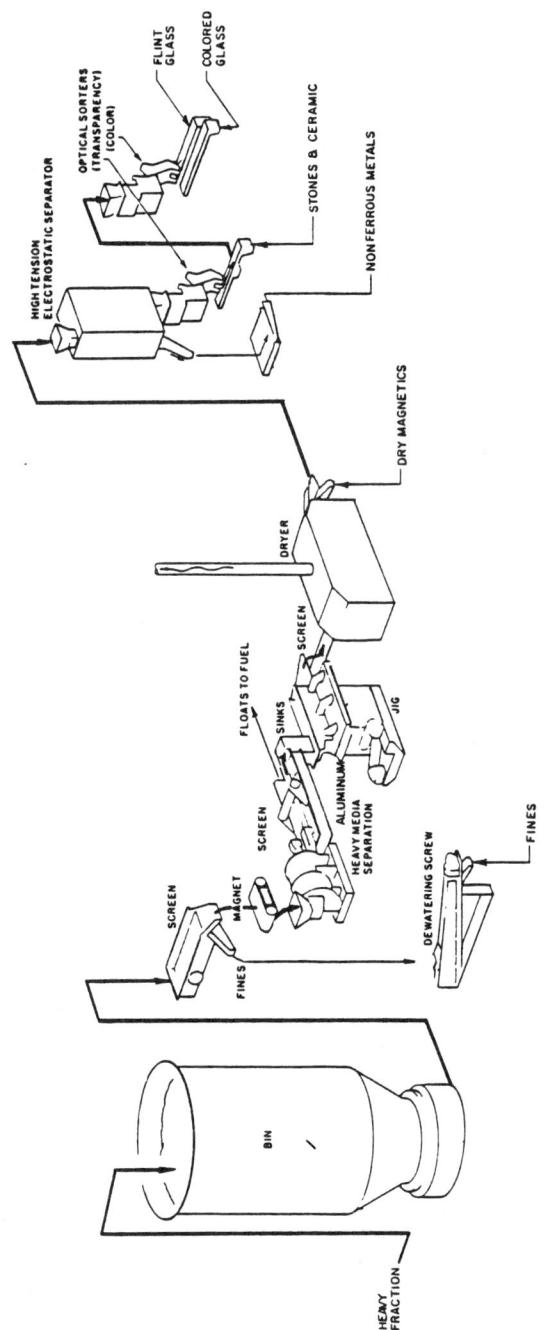

Figure 4-42
Froth Flotation System
(USEPA: Technologies, SW-157.2)

Nonferrous Metal Recovery

Aluminum is the principal nonferrous metal contained in municipal solid waste. It may account for as much as 80 percent of the total nonmagnetic metal content depending, primarily, on the preferred beverage-can package used in an area. The remainder consists of a diverse mixture of copper and zinc alloys. The average amount of aluminum in MSW ranges from 0.5 to 1.0 percent while the other nonferrous metal mixture is about 0.2 to 0.3 percent. Approximately 70 percent of the total aluminum fraction will consist of cans.

Numerous techniques have been tried or are in various stages of development for the recovery of nonferrous metals. Only those which appear to have high potential for commercial development are reviewed here. Recovery of nonferrous metals is only possible when used in conjunction with a ferrous recovery line. Materials rejected from ferrous recovery would be candidate materials for nonferrous recovery.

The operations may be grouped as wet and dry processes, as listed below:

Wet	Dry
Jigging	Eddy Current
Water Elutriation	Electrostatic (high tension)
Heavy Media	

Jigging - Jigging is used to separate materials of different specific gravities. Water is pulsed through a screen causing feed material to separate. The lighter particles are floated to the top while the heavier particles remain near the bottom of the jig. Jigs have been tested for separation of aluminum from heavy nonferrous metals contained in municipal incinerator residue. These tests have produced aluminum fractions of 96 to 98 percent aluminum, while yielding a copper-rich fraction containing copper, lead and zinc. Jigging may be used in conjuction with water elutriation or heavy metal separation.

Water Elutriation - Essentially, this process is classification in a vertical water column; it is used at automobile shredder scrap installations for production of high-grade metal concentrates. In this process, a rising current of water creates an effective specific gravity in excess of that of still water; the effective specific gravity may be controlled at a range of 1.1 to 2.0. Light material flows to the surface and over a weir while heavy material sinks and is picked up by a drum elevator which carries it to a dewatering screen. For processing of MSW, this operation would likely follow air classification and magnetic separation. Remaining organic materials would float over the weir. The sink fraction produced will consist primarily of metals, glass and other inorganic materials.

Heavy Media Separation - This process is another wet technique, from the minerals processing field, which has been tested for use in MSW resource recovery. A finely ground ferro-silicon, magnetite, or other dense material is placed in a water suspension. Constant agitation is applied to maintain the suspension.

The specific gravity of the medium is variable, controlled by the amount of mineral in suspension. For municipal solid waste separation, a specific gravity somewhat less than two may be used to float off rubber and plastics while sinking nonferrous metals, glass, and other inorganic contaminants. A second pass at a specific gravity of about three will float the aluminum and sink the remaining components. Separation in terms of aluminum contained in feedstocks fed to the heavy media separation process is approximately 75 percent while separation of other nonferrous metals is about 85 percent.

Three basic types of heavy media separators are available. They are the drum, cone, and tank type vessels. In addition, a heavy media cyclone (See Figure 4-43) is used where the feed particles are small in size. These systems have been employed on a commercial scale for separation of a number of feedstocks including coal, ores, and shredded automobile scrap.

Eddy Current Separation - This technique is a dry method of separation, based on the principle that nonferrous metals passing through an electromagnetic field will have eddy currents induced within them. These eddy currents produce magnetic fields which repel an applied magnetic field. A repelling force is generated which facilitates separation of the metal. Eddy current separators, also known as traveling-wave separators, employ linear induction motors (LIM) as the traveling-wave source. The LIM is similar to a stator in a typical rotary-induction motor except that it is flat rather than circular. A magnetic wave is made to travel down the length of the LIM. Eddy currents are induced in the nonferrous metals as they move over the LIM causing the generated and induced fields to interact, thus pushing the metals along the field. A separation of the metals is achieved by arranging and phasing the LIM to repel metals away from the feed stream.

Projected recovery of can stock aluminum from pilot eddy current separators is approximately 70 pecent. Preliminary data show that the recovered aluminum may be expected to meet can stock grade aluminum specifications.

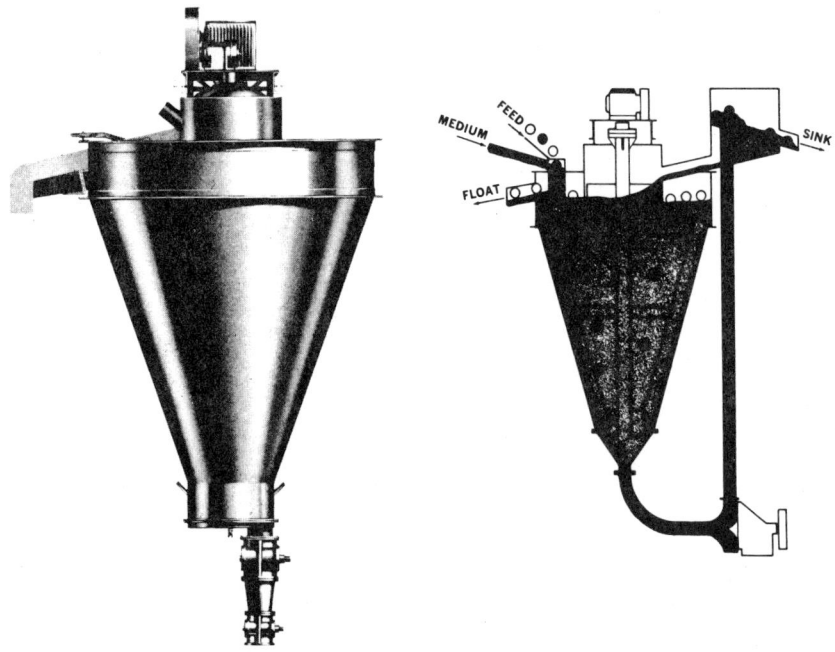

Cone-type Separator

Single-gravity two-product system. Pieces with specific gravity higher than medium sink. Those lower than medium rise.

Figure 4-43
Heavy Media Separator;
WEMCO Div. of Envirotech
(Reference 13)

Electrostatic Separation - This procedure is another dry system for nonferrous metal separation. Separation is achieved by making use of the electrical conductivity of the particles in the feed stream. As feed material enters the electrostatic field, particles become charged. Conductors immediately lose their charge and fall from the drum while nonconductors retain a surface charge and are pinned to the ground drum. They are swept off once they have passed the splitter point. These units have been used for mineral separation and food processing. (See Figure 4-44). Test results with electrostatic separation indicate approximately 80 percent of aluminum is recovered with purities of up to 88 percent.

MATERIALS HANDLING UNIT PROCESS

The following materials handling processes are commonly used at resource recovery facilities.
- Apron conveyors
- Belt conveyors
- Pneumatic conveyors
- RDF storage systems

Apron Conveyors

Apron conveyors consist of a series of jointed steel pans connected by and driven by chains. Apron conveyors are used to convey unprocessed solid waste, generally as infeed to a primary shredder or trommel.

Many types of apron conveyors are available including piano-hinge, Z-pan, double beaded, and inboard and outboard roller. The merits of these designs are a subject of dispute among the various manufacturers.

One type of apron conveyor which has proven successful in handling solid waste is the double beaded outboard roller type shown in Figure 4-45. Figure 4-46 shows a selection graph for conveyor width: as indicated, conveyor speed should be kept between 10 an 30 fpm and incline less than 40°. Conveyor width should always be equal to or slightly less than the throat width of the next piece of equipment. If the feeding conveyor is wider than the subsequent conveyor "funneling" is very likely to occur causing bridging and blockages. Changes in direction should not be attempted when conveying unshredded solid waste.

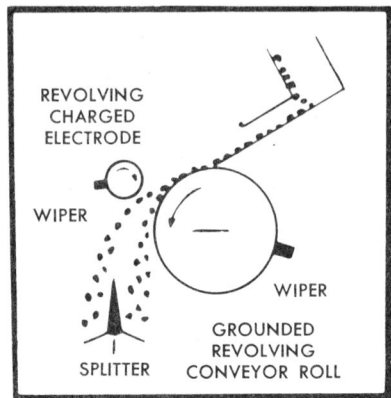

Charged electrode selectively pulls material beyond flow splitter.

Principle of high voltage electrostatic separation; Dings Co.

Figure 4-44
High Voltage Electrostatic Separator;
Dings, Inc.
(Reference 13)

Unit Processes of Resource Recovery 119

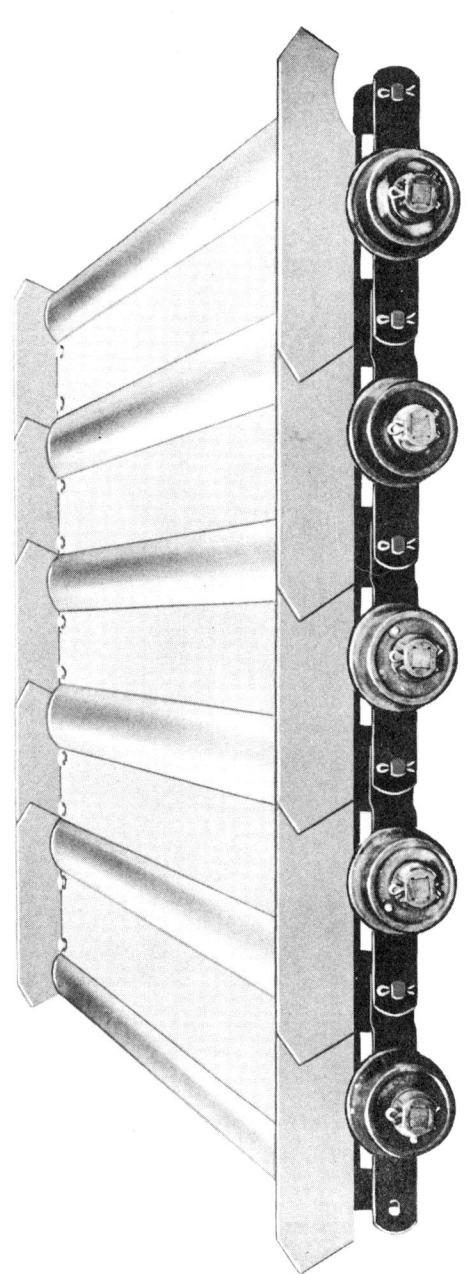

Figure 4-45
Outboard roller apron conveyor
(Rexnord)

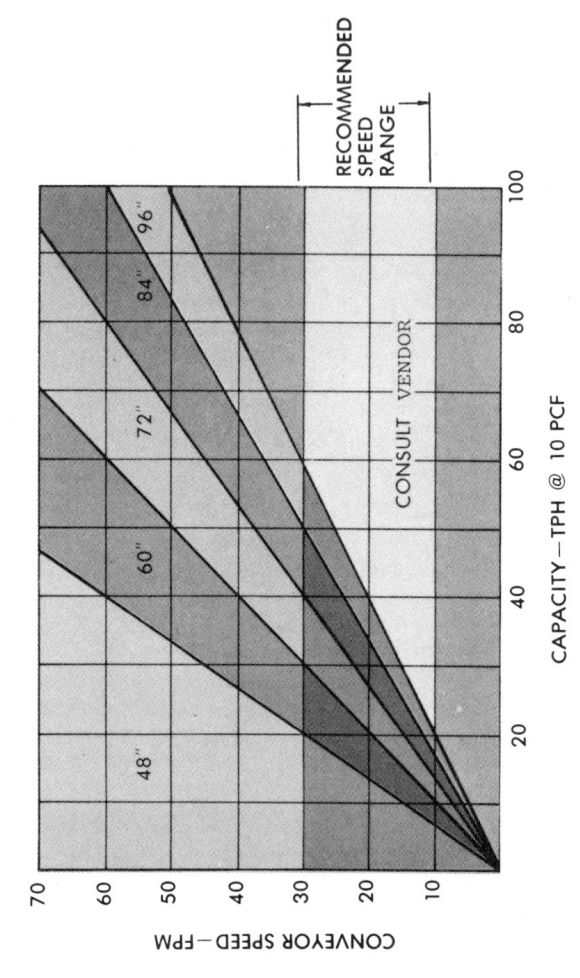

Figure 4-46
(Rexnord)

Figures 4-47 and 4-48 show a typical shredder infeed installation. The use of separate horizontal and inclined conveyors should be noted. Use of this arrangement helps to avoid the overloading of the inclined conveyor due to the tendency of solid waste to move in mass up the incline.

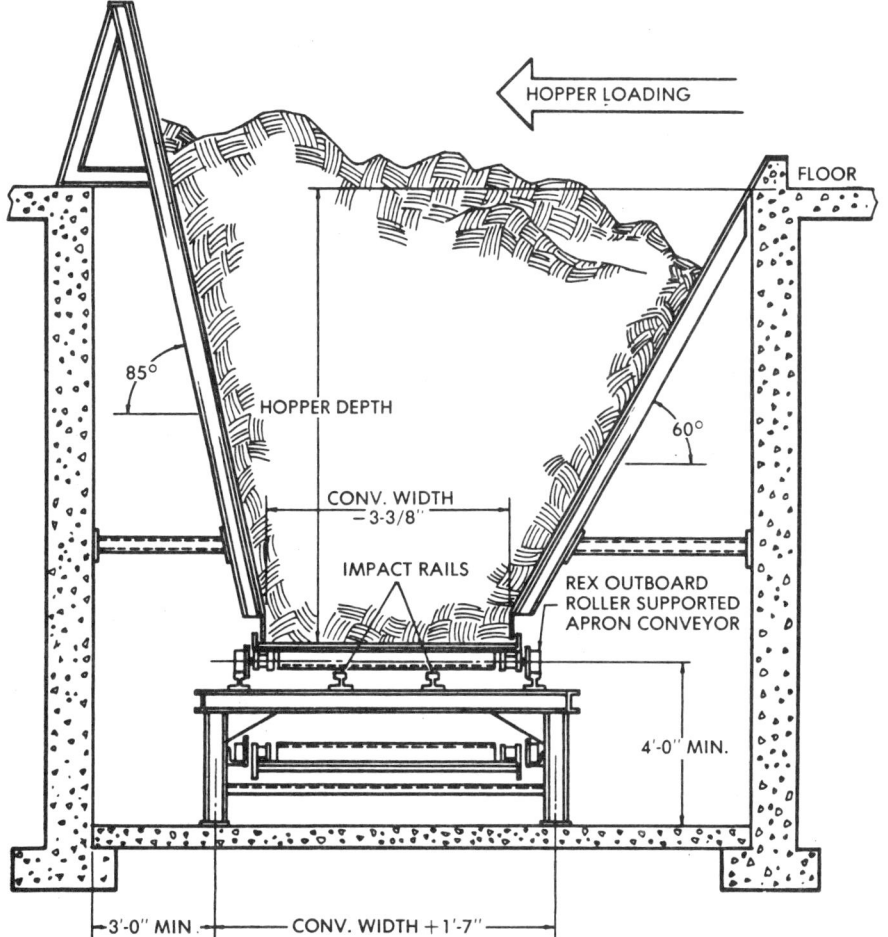

TYPICAL SECTION THRU RECEIVING HOPPER

Figure 4-47
(Rexnord)

Belt Conveyors

Belt conveyors are widely used in resource recovery systems to convey solid waste and various waste components

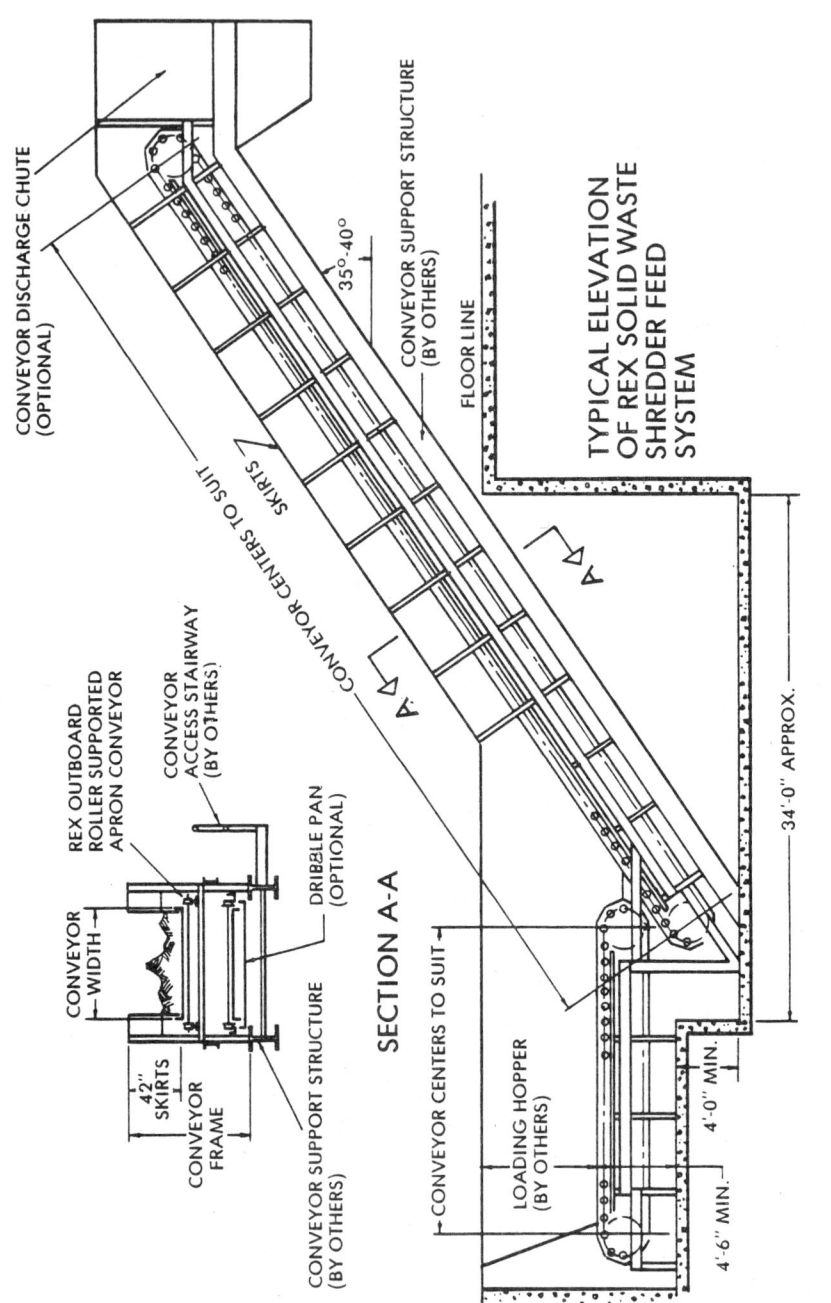

Figure 4-48
(Rexnord)

after shredding. One exception is the shredder discharge conveyor itself: because material exits the shredder at high temperature and/or velocity, metal apron or vibrating conveyors are advisable.

Preliminary selection of a belt conveyor can be made based on Tables 4-6, 4-7, and 4-8, using the following rules:

o Maximum Recommended Speeds:

    Shredded Material - 350 fpm
    Light Fraction   - 350 fpm
    Heavy Fraction   - 400 fpm

o Use 35° trough angle

TABLE 4-6

ANGLE OF SURCHARGE—ANGLE OF REPOSE

5° Angle of surcharge	10° Angle of surcharge	20° Angle of surcharge
0°-20° Angle of Repose	20°-30° Angle of Repose	30°-35° Angle of Repose
Fluffed, Shredded or Light Fraction	Light Fraction or Heavy Fraction	Heavy Fraction Glass, Stones Dirt, etc.
Density: — 3 - 5 lbs./cu.ft.	Density: — 5 - 10 lbs./cu.ft.	Density: — over 10 lbs./cu.ft.

(REXNORD)

TABLE 4-7

MAXIMUM IDLER SPACING IN FEET.

BELT WIDTH	Carrying Idlers Weight of Material Conveyed in pounds per cubic foot.				RETURN IDLERS	AVERAGE BELT WEIGHT Lbs./ft.
	15	30	50	100		
24	5.0	5.0	4.5	4.0	10.0	4.5
30	5.0	5.0	4.5	4.0	10.0	6.0
36	5.0	5.0	4.5	4.0	10.0	9.0
42	4.5	4.5	4.5	3.5	10.0	11.0
48	4.5	4.5	4.0	3.5	10.0	14.0
54	4.5	4.5	4.0	3.5	10.0	16.0
60	4.0	4.0	3.0	3.0	10.0	18.0
72	4.0	4.0	3.0		8.0	20.0
84	3.5	3.0	2.0		8.0	23.0
96	3.0	3.0			6.0	26.0

(REXNORD)

TABLE 4-8

35° TROUGHING

Belt Width	Surcharge Angle Degrees	Cross Sectional Area Square Feet	CAPACITY — CUBIC FEET PER HOUR BELT SPEED — FEET PER MINUTE							Maximum Lump Size Inches
			100	150	200	250	300	350	400	
24	0	.222	1332	1998	2664	3330	3996	4662	5328	5
	5	.247	1482	2223	2964	3705	4446	5187	5928	
	10	.273	1637	2456	3274	4092	4910	5730	6547	
	15	.298	1790	2685	3580	4475	5371	6265	7162	
	20	.325	1949	2924	3898	4872	5846	6822	7795	
30	0	.364	2184	3276	4368	5460	6552	7644	8736	6
	5	.405	2430	3645	4860	6075	7290	8505	9720	
	10	.446	2674	4011	5347	6685	8020	9359	10694	
	15	.487	2923	4384	5846	7308	8770	10230	11693	
	20	.530	3178	4767	6355	7945	9533	11123	12710	
36	0	.501	3006	4509	6012	7515	9018	10521	12024	7
	5	.601	3606	5409	7212	9015	10818	12621	14424	
	10	.661	3965	5948	7930	9912	11894	13878	15859	
	15	.722	4334	6501	8669	10835	13003	15169	17338	
	20	.784	4704	7056	9408	11760	14112	16464	18816	
42	0	.752	4512	6768	9024	11280	13536	15792	18048	8
	5	.835	5010	7515	10020	12525	15030	17535	20040	
	10	.918	5510	8265	11020	13775	16531	19285	22042	
	15	1.003	6019	9028	12038	15048	18058	21066	24077	
	20	1.089	6533	9800	13065	16332	19598	22866	26131	
48	0	.998	5988	8982	11976	14970	17964	20958	23952	10
	5	1.108	6648	9972	13296	16620	19944	23268	26592	
	10	1.218	7310	10965	14620	18275	21931	25585	29242	
	15	1.330	7978	11967	15955	19945	23933	27923	31910	
	20	1.443	8659	12988	17318	21648	25978	30306	34637	
54	0	1.279	7674	11511	15348	19185	23022	26859	30696	11
	5	1.419	8514	12771	17028	21285	25542	29799	34056	
	10	1.560	9360	14040	18720	23400	28080	32760	37440	
	15	1.702	10214	15321	20429	25535	30643	35749	40858	
	20	1.847	11083	16625	22166	27708	33250	38790	44333	
60	0	1.595	9570	14355	19140	23925	28710	33495	38280	12
	5	1.769	10614	15921	21228	26535	31842	37149	42456	
	10	1.943	11659	17488	23318	29148	34978	40806	46637	
	15	2.121	12725	19088	25450	31812	38174	44538	50899	
	20	2.301	13805	20708	27610	34512	41414	48318	55219	
72	0	2.330	13980	20970	27960	34950	41940	48930	55920	14
	5	2.583	15498	23247	30996	38745	46494	54243	61992	
	10	2.838	17025	25538	34051	42562	51077	59588	68102	
	15	3.095	18571	27856	37142	46428	55714	64998	74285	
	20	3.357	20140	30210	40282	50350	60422	70490	80563	
84	0	3.287	19722	29583	39444	49305	59166	69027	78888	16
	5	3.671	22026	33039	44052	55065	66078	77091	88104	
	10	4.058	24345	36518	48691	60862	77037	85208	97382	
	15	4.449	26693	40040	53385	66732	80078	93426	106771	
	20	4.847	29083	43624	58166	72708	87250	101790	116333	
96	0	3.861	23166	34749	46332	57915	59498	81081	92664	18
	5	4.376	26256	39384	52512	65640	78768	91896	105024	
	10	4.894	29366	44049	58733	73415	88099	102781	117465	
	15	5.419	32515	48772	65030	81288	97545	113802	130060	
	20	5.954	35722	53583	71443	89305	107165	125027	142886	

(REXNORD)

Example 4-5 illustrates the selection of a belt conveyor.

EXAMPLE 4-5

Select a belt conveyor to handle 50 tph of shredded solid waste.

1. From Table 4-6, angle of surcharge is 5°, density is 3-5 lbs/cu ft.
2. Calculate volumetric flow rate

$$\text{Cu ft/hr} = \frac{\text{tph} \times 2000}{\text{density}} = \frac{50 \times 2000}{3} = 33{,}333$$

3. From Table 4-8, given a belt speed of 350 fpm and a surcharge angle of 5° select minimum belt width.

   Belt width = 60 inches
   Capacity = 37,149 cu ft/hr

4. From Table 4-7, select idler spacing

   Idler spacing = 4 ft
   Return idler spacing = 10 ft

Pneumatic Conveyors

Pneumatic conveying systems for the transport of the light fraction of shredded solid waste (RDF) are available. These systems usually convey the material from an air classifier to storage and consist of a blower, connecting ductwork, a cyclone for separating the light fraction from the conveying air, and a rotary airlock to discharge the light fraction from the cyclone. Pneumatic conveyors can also be also used to move shredded material from storage to an incinerator.

Typical air velocities to convey various materials are shown in Table 4-9. It should be noted that abrasion by entrained glass fines present potential wear problems in these systems.

TABLE 4-9

TYPICAL AIR VELOCITIES TO CONVEY VARIOUS MATERIALS

Material	Air Velocity (ft./min.)
Grain dust	2,000
Wood chips and shavings	3,000
Sawdust	2,000
Jute dust	2,000
Lint	1,500
Rubber dust	2,000
Metal dust (grinding)	2,200
Lead dust	5,000
Brass turnings (fine)	4,000
Fine coal	4,000

Source: Reference 4.

RDF Storage Systems

Various systems available for the storage and retrieval of RDF are discussed below:

Atlas Silo - The Atlas system, shown in Figure 4-49 consists of a flat concrete floor upon which the RDF is stacked, a metal silo which encloses the RDF, and a system of peripheral sweep bucket chains which reclaim the material from the edge of the pile.

The Atlas system is one of the first tried, having been installed at East Hamilton, Ontario in 1972. Other resource recovery facilities using the Atlas system include Ames, Iowa; Milwaukee, WI; and the Chicago Southwest facility. Problems with this system have included bridging in the silo, excessive wear of the support floor, and difficulty in controlling fires.

Live Bottom Bins - A number of manufacturers market fabricated steel bins for RDF storage. Figure 4-50 shows one such bin which uses screw conveyors for RDF retrieval. Bins of this type are expensive for large scale RDF storage but are useful to contain short-term surges.

Crane and Bucket Retrieval - The Black Clawson Hempstead facility uses the following system for RDF storage:
- o RDF is dumped on the floor of a large enclosed room by a system of distributing conveyors.
- o RDF is reclaimed by an overhead bridge crane.

No problems with this system were reported during the time the Hempstead facility was in operation.

Parascrew system - The Akron and Hooker facilities use a unique system shown in Figure 4-51. RDF is distributed within a storage building by overhead shuttle conveyors and to reclaimed by hydraulicly driven auger screw conveyors.

Unit Processes of Resource Recovery 127

ATLAS T2 AND T3 TYPE STORAGE
Figure 4-49

Figure 4-50
Live Center Bin
(Sprout-Waldron)

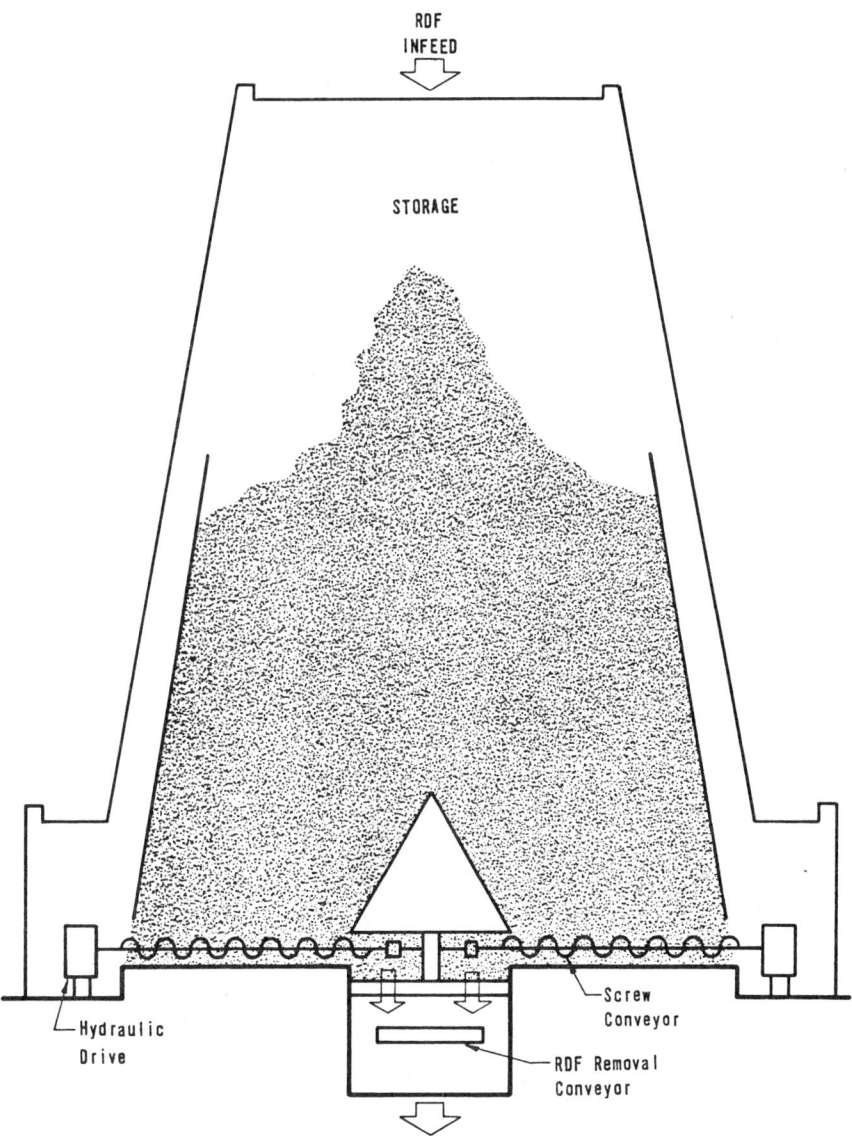

Figure 4-51
PARASCREW RDF STORAGE SYSTEM

REFERENCES

1. Howe Richardson Scale Company, Weight Tickets Type A for M-R Printer, Accutronic Truck Scales, Super Cast Truck Scales.
2. General Electric, Solid Waste Management Technology Assessment, Van Nostrand Reinhold Co., 1975.
3. Fairbanks Weighing Division of Colt Industries, Scales for Solid Waste Management.
4. Hill, Roger M., "Solid Waste-Ring in the Old" Waste Age, September 1977.
5. "Why Start with Shredding," American City and County October 1978.
6. Midwest Research Institute, Size Reduction Equipment for Municipal Solid Waste, National Technical Information Service, 1973.
7. USEPA, Solid Waste Shredding and Shredder Silicon, November 1974.
8. Hammermills Inc., Bulldog Refuse Shredders.
9. Tchobanoglous, et al, Solid Wastes, Engineering Principals and Management Issues, McGraw-Hill, 1977.
10. Combustion Engineering, Inc., Generating Steam From Refuse in Industrial Boilers, International District Heat Association, June 1979.
11. Rochford, R.S. and S.J. Witkowski, Babcock & Wilcox, Considerations in the Design of a Shredded Municipal Refuse Burning and Heat Recovery System, ASME/Eighth National Waste Processing Conference, 1978.
12. National Center for Resource Recovery, New Orleans Resource Recovery Facility Implementation Study: Equipment, Economics, Environment, September 1977.
13. National Center for Resource Recovery, Materials Recovery System, Engineering Feasibility Study, December 1972.
14. USEPA, Resource Recovery Plant Implementation: Guides for Municipal Officials—TECHNOLOGIES, 1976.
15. Brunner, C.R. Handbook of Incinerator Systems, Van Nostrand Reinhold Co., 1983.

5

Resource Recovery Systems

INTRODUCTION

In this chapter the unit processes discussed in the previous section will be assembled into complete resource recovery systems.

Resource recovery systems considered in this chapter are subdivided into the following categories:

- o Mass burning systems

- o Refuse derived fuel (RDF) systems
  - RDF systems without burning
  - RDF systems with burning

- o Pyrolysis systems

- o Starved air incineration systems

EUROPEAN VS. AMERICAN PROJECTS

In Europe, after the Second World War, the disposal of municipal solid waste at a central location, by incineration, was given impetus by the following factors:

- o Population concentrations and increases required the use of more land for housing and for farming. The use of land for burying refuse was becoming impractical.

- o Technology developed to the point where it became economical to generate energy, i.e., steam and/or electric power, from incineration. Economics of scale dictated that the larger the facilities the more efficient would be its energy generation potential.

- o In general all utilities, including refuse disposal and electric/steam power generating industries, were state owned. The interests of the electric utility and the refuse disposal authority were therefore, common. This conflicts with conditions in the United States where refuse collection is a public or governmental function and electric power generation is generally a private sector function. Cooperation between these agencies in Europe has promoted the development of energy producing incinerators. The power utility readily purchases energy from an incineration facility providing revenues for the incinerator operation.

- o The higher cost of fossil fuel, particularly fuel oil, has helped promote energy generation, hence, central disposal facilities.

Only in the past decade, when the United States came to the realization that the cost and availability of energy was unreliable and out of its control, was a serious attempt made to generate energy from waste in central collection and incineration facilities.

MASS BURNING
General

Mass burning is defined as the combustion of solid waste as received from the collection vehicle. The only processing that occurs is the blending of wastes, and the removal of unprocessable or hazardous items including white goods such as refrigerators, stoves, working machines, etc., both of which are accomplished in the refuse storage pit by the crane operator.

Mass burning systems avoid the capital and operating costs associated with refuse processing. However, they must deal with raw refuse, which is a poor fuel at best. In addition, mass burning systems forego opportunities for materials recovery, with the exception of ferrous metals recovery from the ash. This "back-end" recovered product is worth less than "front-end" (before combustion) recovered metals and is often unsaleable because of its degradation as it passes through the burning zones of the incinerator.

Process

Most mass burning systems use water tube wall (often called "waterwall") furnaces for combustion. These furnaces have largely replaced the older refractory-lined, stoker-fired incinerators which were used for refuse incineration in the past. Waterwall units are enclosed by closely spaced water filled steel tubes (see Fig. 4-29) which recover heat by both radiation and convection. This heat recovery enables them to operate at lower temperatures than refractory lined units which, in turn, makes them smaller and cheaper to build, and more efficient in energy recovery.

A typical mass burning system is shown in Figure 5-1. Refuse is dumped into a storage pit and then loaded into refuse hoppers by crane. The refuse falls into the furnace by gravity and is moved through the furnace by a system of agitating grates. Unburned residue falls off the end of the grates for land disposal.

The combustion gases, after giving up heat to various boiler sections, are cleaned by electrostatic precipitators and then discharged to the atmosphere.

Because of the nature of solid waste, it is difficult to achieve good mixing with air as required for complete combustion. A variety of proprietary grate systems have been used to achieve this mixing, as described in Chapter 4.

Status

Mass burning incineration is the most proven of energy recovery technologies, having been in operation in Europe for over 20 years. The earliest US installation, a 750 tpd system in Oceanside, NY, has been in operation since 1965. Nine facilities of this type are in operation today in the US, as indicated below:

Facility	Capacity Tons/Day	Year Operational
Oceanside, NY	750	1965
Norfolk, VA	360	1967
Chicago N.W., IL	1600	1971
Braintree, MA	250	1971
Harrisburg, PA	720	1972
Nashville, TN	400	1974
Saugus, MA	1200	1975
Portsmouth, NH	160	1976
Hampton, VA	200	1980

RDF SYSTEMS

General

The goal of RDF systems is to improve the combustion of solid waste by processing to produce a fuel of lower moisture

134 Energy and Resource Recovery from Waste

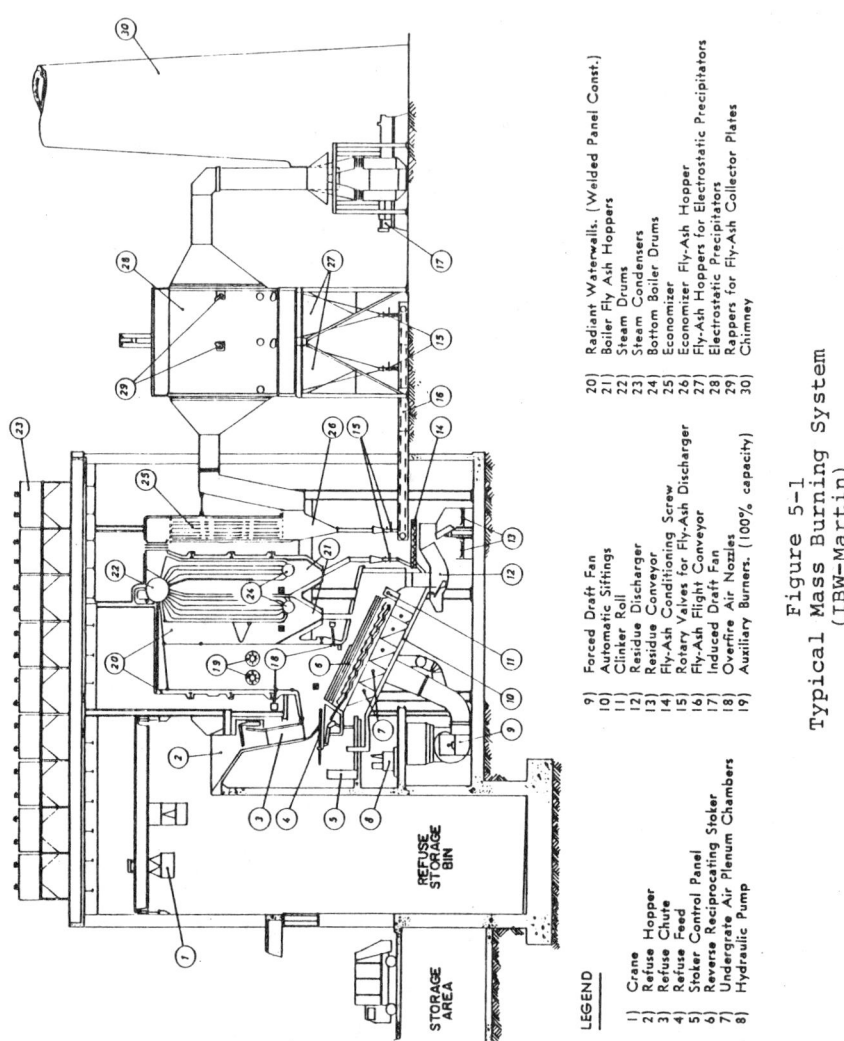

LEGEND

1) Crane
2) Refuse Hopper
3) Refuse Chute
4) Refuse Feed
5) Stoker Control Panel
6) Reverse Reciprocating Stoker
7) Undergrate Air Plenum Chambers
8) Hydraulic Pump
9) Forced Draft Fan
10) Automatic Siftings
11) Clinker Roll
12) Residue Discharger
13) Residue Conveyor
14) Fly-Ash Conditioning Screw
15) Rotary Valves for Fly-Ash Discharger
16) Fly-Ash Flight Conveyor
17) Induced Draft Fan
18) Overfire Air Nozzles
19) Auxiliary Burners. (100% capacity)
20) Radiant Waterwalls. (Welded Panel Const.)
21) Boiler Fly Ash Hoppers
22) Steam Drums
23) Steam Condensers
24) Bottom Boiler Drums
25) Economizer
26) Economizer Fly-Ash Hopper
27) Fly-Ash Hoppers for Electrostatic Precipitators
28) Electrostatic Precipitators
29) Rappers for Fly-Ash Collector Plates
30) Chimney

Figure 5-1
Typical Mass Burning System
(IBW-Martin)

content, more uniform size, and lower ash content. Some RDF systems attempt to produce a fuel good enough to sell as a supplement to coal or oil in utility or large industrial boilers. (Producing a salable fuel obviates the need for investing in boilers at the resource recovery facility.) Other RDF systems use a "dedicated boiler" (a boiler designed for RDF use on site) but process the MSW to RDF to improve combustion efficiency and reduce boiler maintenance.

Fluff RDF

Fluff or crude RDF systems are the most common RDF systems. Generally these systems include shredders, air classifiers, and other materials recovery and separation subsystems as required to achieve the desired RDF product specification.

Two basic fluff RDF variations exist. One type is aimed at producing a supplemental fuel for use in coal fired utility boilers. Examples include the St. Louis-Union Electric project; Ames, Iowa; the Americology facility in Milwaukee; and the Chicago Southwest facility. Figure 5-2 illustrates the Americology process. This process recovers aluminum, glass, ferrous metals, newspaper and corrugated, and a fuel product with heating value of about 5000 Btu per lb as received.

A comparison of an RDF of this type and coal is given below:

	St. Louis RDF	Bituminous Coal
Moisture (%)	27.8	2.6
Sulfur (%)	0.15	1.3
Chloride (%)	0.34	NA
Ash (%)	21.6	9.1
Btu/lb	4,631	13,610
Minimum Ash softening temperature (°F)	1,880	1,940

Clearly, RDF is an inferior fuel to coal in every characteristic except sulfur.

The second type of fluff RDF system is aimed at producing a fuel for use in a dedicated boiler, usually of the semi-suspension, spreader stoker type. Since the system does not aim at selling the fuel RDF quality is less important and simplification in processing and attendant cost savings can be achieved. Offsetting this process cost savings to some extent is the need to invest in specialty combustion equipment, avoided if the fuel is sold.

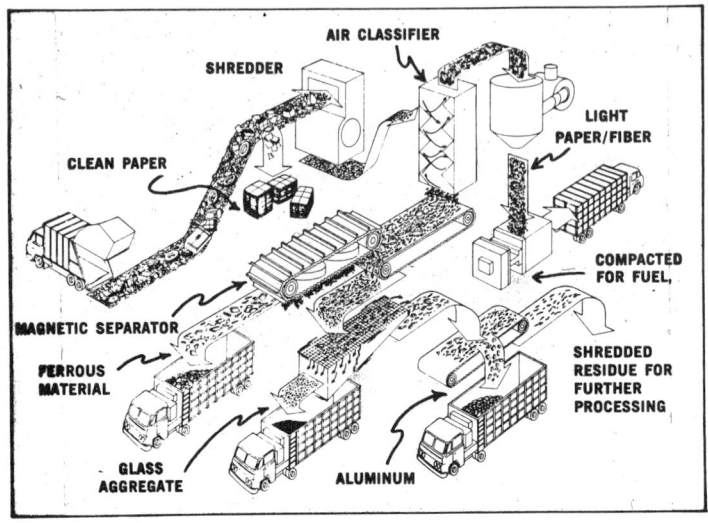

Figure 5-2
Americology Process
(American Can Company)

Figure 5-3 illustrates a typical RDF system of this kind. Its relative simplicity, particularly the absence of two-stage shredding, should be noted.

Examples of this type of system include Hamilton, Ontario (SWARU); Akron, OH; and the Hooker Chemical Facility in Niagara Falls, NY.

Densified RDF (d-RDF)

Some of the disadvantages of fluff RDF as a fuel are related to its low density, high transport and storage costs, and difficulties in retrieval from storage. Densified RDF processes attempt to address these problems by forming the RDF into higher density pellets. Pellet mills (see Figure 5-4) and other densifying equipment have been used with some success to increase RDF densities from an initial 3 to 5 lbs/cu ft to as much as 50 lbs/ cu ft.

Densified RDF systems are in the research and development stage. Attempts at producing densified RDF on a large scale (Americology) have not met with success.

Resource Recovery Systems 137

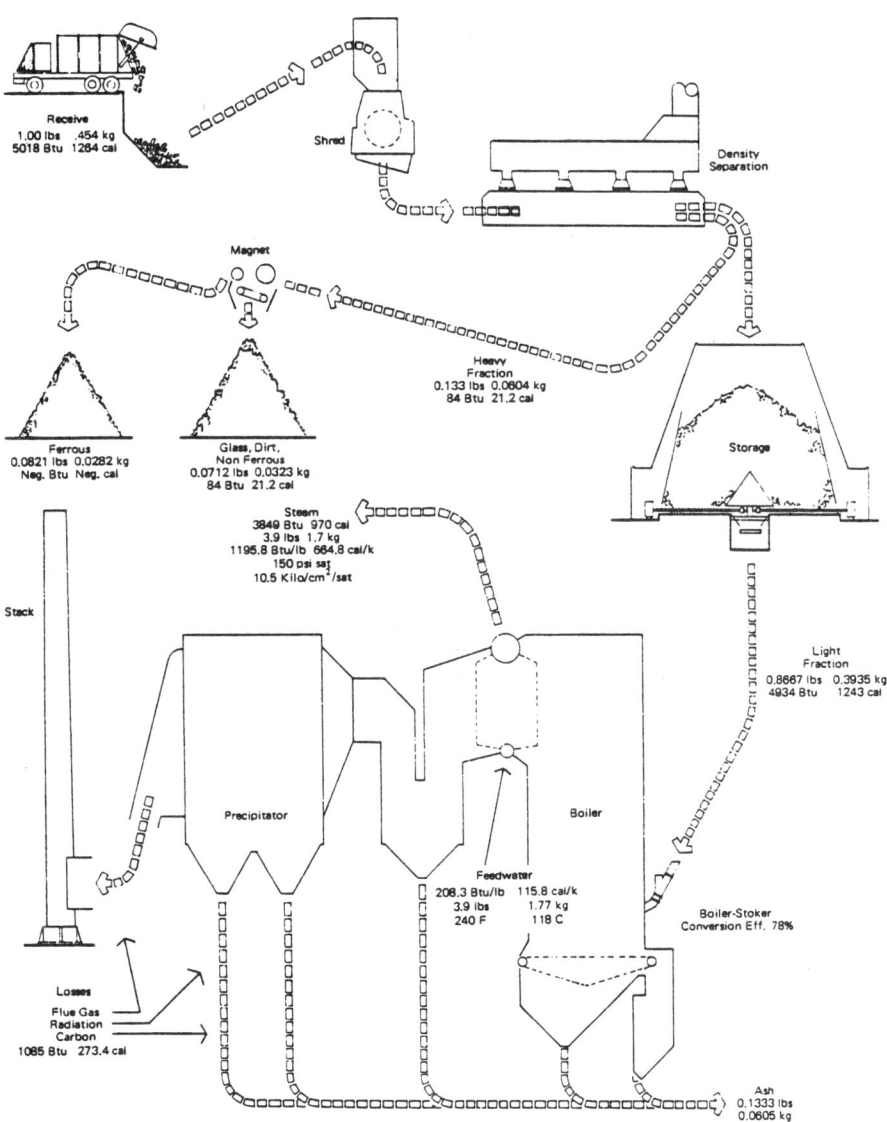

Figure 5-3
Typical Fluff RDF System
(Reference 3)

Densifier

In operation, the material to be pelleted is fed continuously to the pelleting chamber. Here it is directed equally into the two areas formed by the steel rollers and the inside face of the die. Rotation of the die in contact with the rollers causes rollers to turn. The material is thus compressed and, under extreme pressure, forced through the die holes.

Figure 5-4
Pellet Mill
(California Pellet Mill Co.)

Dust RDF

A variation of the RDF system is the ECO-Fuel II process developed by Combustion Equipment Associates (CEA). The CEA process produces a dust-like fuel resembling powdered coal with approximately the following properties:

PROPERTIES OF ECO-FUEL II

Estimated Chemical Analysis

	Percent by Weight (as fired)
Carbon	41.6 - 47.3
Hydrogen	5.5 - 6.3
Oxygen	33.9 - 38.6
Nitrogen	0.6 - 1.5
Ash	5.0 - 12.0
Sulfur	0.1 - 0.6
Chloride	0.1 - 0.7
Water	1.0 - 5.0

Generalized Properties

	Percent by Weight (as fired)
Combustible	88.6
Ash	9.4
Moisture	2.0
	100.0

Higher Heating Value	7800 Btu/lb
Particle Size	<0.015 inch
Bulk Density	30-35 lbs/CF
Storage Life	Indefinite

The ECO-Fuel production process is illustrated in Figure 5-5. This process is different from the production of ordinary RDF in several respects:

- o A flail mill is substituted for the hammermill type primary shredder saving substantial horsepower.

- o A proprietary chemical embrittling agent is used to facilitate decomposition of cellulose fibers.

- o A hot ball mill is used to pulverize the embrittled coarse RDF.

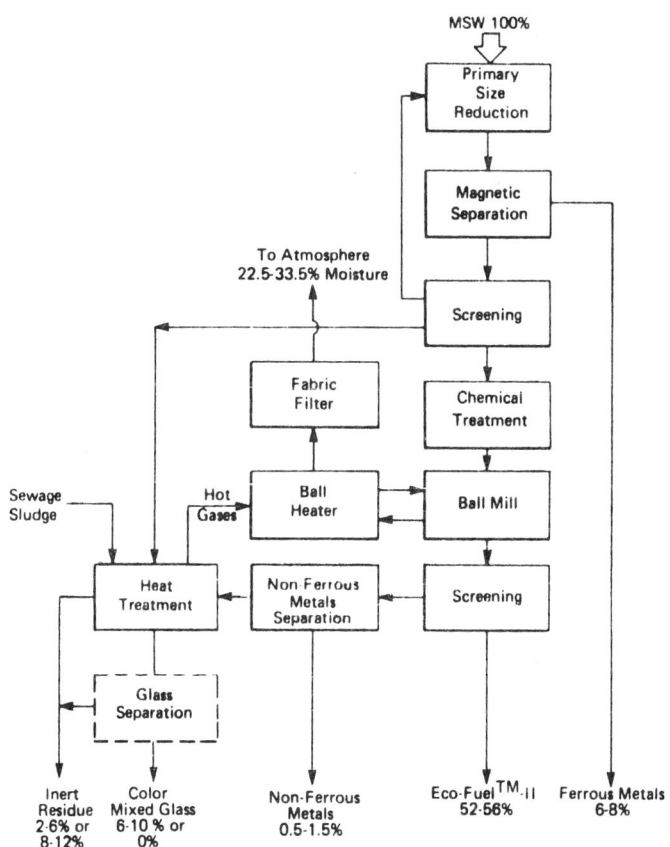

Figure 5-5
ECO-Fuel Production Process
(Combustion Equipment Associates
and Arthur D. Little, Inc.)

The advantages of the CEA process include a more saleable, more storable fuel; the disadvantages include higher processing costs.

CEA has built a 2000 tpd facility in Bridgeport, CT. This facility was undergoing startup in December 1980 when CEA entered bankruptcy proceedings: this facility is now closed.

Wet-Pulped RDF

RDF can also be produced by a wet processing system where the waste is converted into a slurry. In this proprietory process, a "hydropulper" is used in place of a shredder. The hydropulper, which resembles an electric food blender, reduces the size of the refuse particles, and ejects unshreddable items such as large pieces of metal. Water is mixed with the refuse in the hydropulper producing a slurry which is then processed to separate glass, aluminum, and ferrous metal. This system produces an RDF with a moisture content of about 50 percent (about double the moisture content of raw refuse) and a heating value of 4,100 Btu/lb. This RDF can only be burned on a grate, not in suspension.

The wet-pulp RDF process has the advantage of producing a uniform, easily handled RDF, and of eliminating dust and shredder explosion hazards. Its disadvantages include the low energy of the RDF produced, and the cost and mechanical complexity of the processing system.

This system was developed by the Black Clawson Co. The only completed facility using this technology is located at Hempstead, NY.

A diagram of the Hempstead facility is shown in Figure 5-6.

PYROLYSIS SYSTEMS

General

Several pyrolysis systems for processing solid waste have been proposed, tested at pilot and/or demonstration scale, and in one case demonstrated at full scale. These systems include the Union Carbide "Purox" process, the Monsanto Landguard System, the Occidental Petroleum flash pyrolysis system, and others. None of the above mentioned systems is being marketed at this time.

Resource Recovery Systems   141

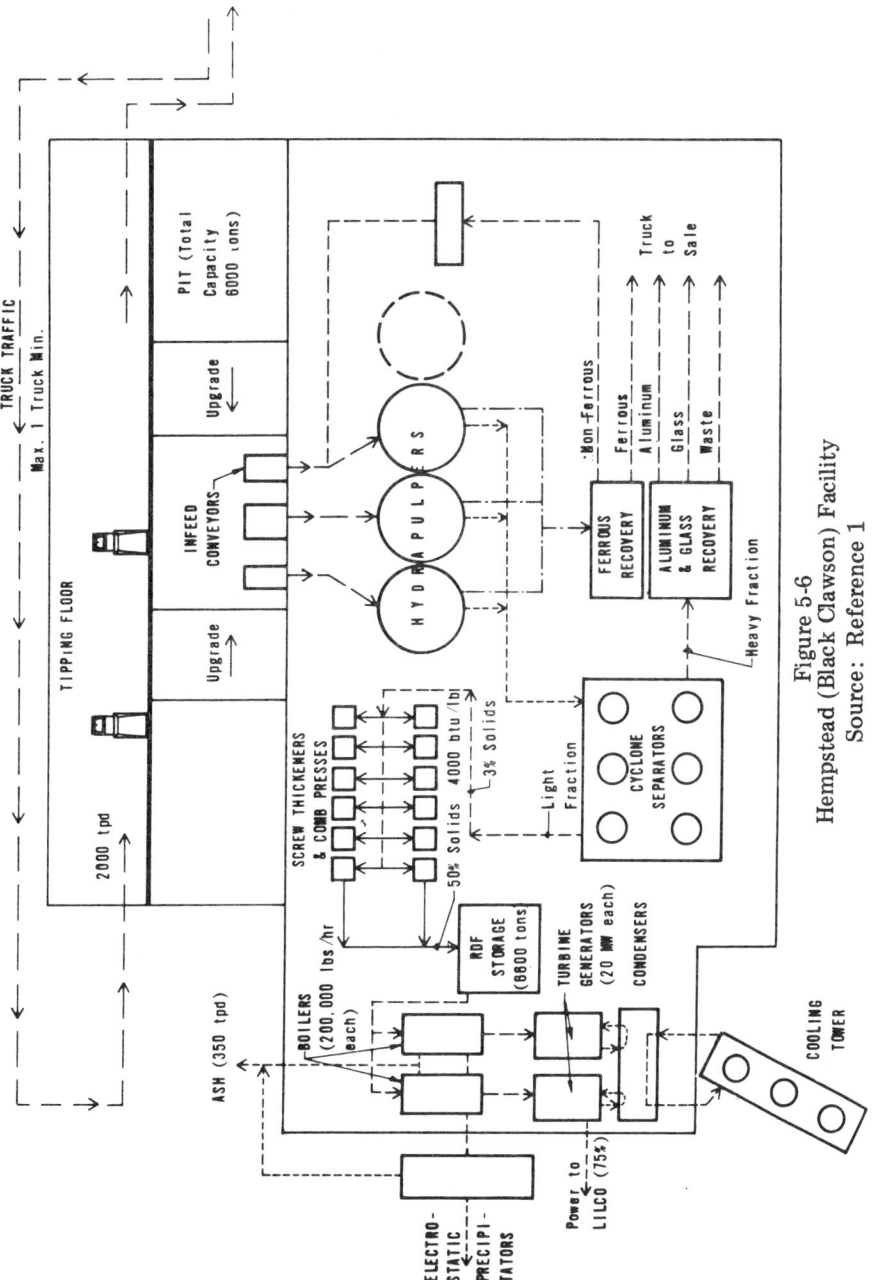

Figure 5-6
Hempstead (Black Clawson) Facility
Source: Reference 1

The balance of this section will address the Andco-Torrax pyrolysis system, and the Union Carbide "Purox" system. The Monsanto system is no longer being marketed.

Andco-Torrax

The Andco-Torrax system is an air fed slagging pyrolysis system. Unshredded refuse is fed into the top of a vertical shaft furnace (gasifier) and descends by gravity through three process zones - drying, pyrolysis and combustion. The function of the drying zone is to evaporate the moisture in the refuse and to act as a plug to restrict the in-flow of air during charging. The drying zone extends downward below the bottom of the lantern section of the gasifier. The lantern is the point where the combustible gas and vapor mixture from the pyrolysis process is drawn out under negative pressure.

Below the drying zone is the pyrolysis zone where the dried refuse is thermally decomposed to a residual mixture of carbon and inerts, releasing a mixture of combustible gases and vapors.

The heat for pyrolyzing and drying the refuse is supplied by the partial combustion of the carbon char with preheated air supplied at the base of the gasifier in the primary combustion zone. The heat generated by this combustion process also transforms the non-combustible materials to a molten slag. The molten slag is drained continuously through a sealed slag tap into a water quench tank to produce a black, glassy aggregate which contains no carbon or putrescible material.

The Andco-Torrax system produces a fuel gas consisting primarily of CO, CO_2, hydrocarbon vapors, hydrogen, nitrogen and water vapor. The heating value of the gas is 120-150 Btu per cu ft. This gas is fired immediately after gasification in a secondary combustion chamber. Energy recovery is accomplished with a waste heat boiler after the combustion chamber (see Figure 5-7).

Andco has several installations in Europe (see Figure 5-8) and Japan; one installation is currently under construction in the US, at Walt Disney World in Florida.

Union Carbide Purox System

The Union Carbide Purox System is an oxygen-fed slagging

Resource Recovery Systems 143

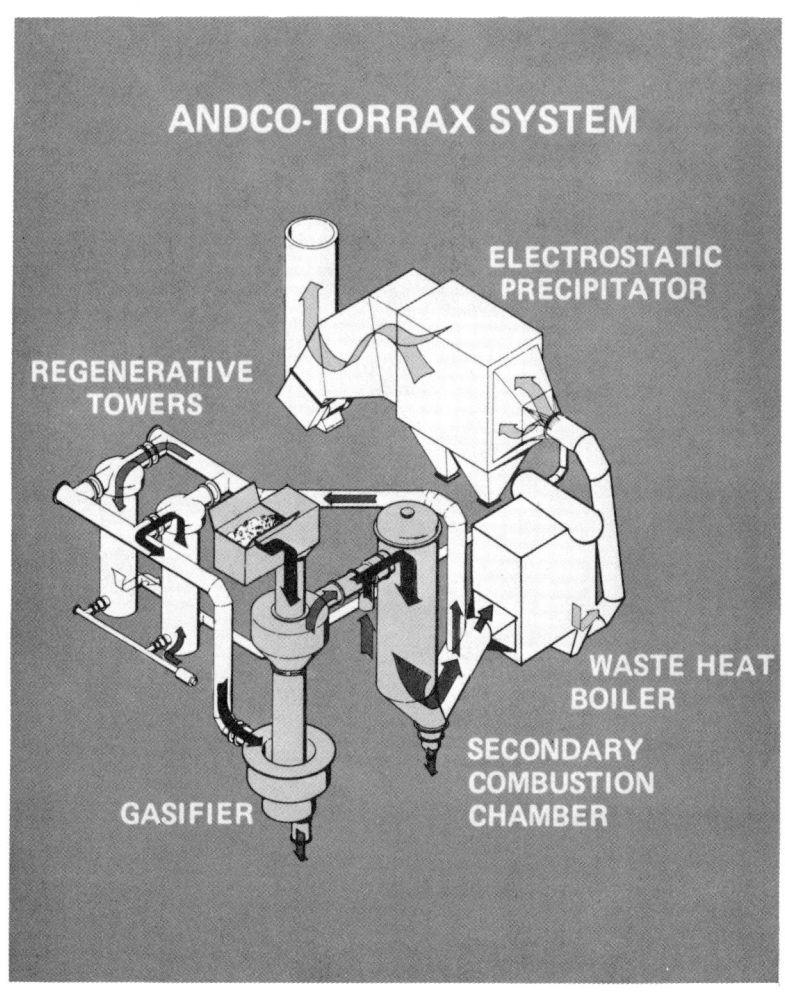

Figure 5-7
(Andco Inc.)

Figure 5-8
Commercial Andco-Torrax
System in Frankfurt, West Germany
(Andco Inc.)

pyrolysis system. Shredded refuse with ferrous metals removed are loaded into the top of a vertical shaft furnace and go through the same drying, pyrolysis, combustion process as in the Andco-Torrax gasifier. The process produces a molten slag (called frit) and a medium Btu fuel gas meeting the following specifications:

TYPICAL PUROX SYSTEM FUEL GAS ANALYSIS

COMPOSITION (Dry Basis)	Volume %
H_2	26
CO	40
CH_4	5
$C_2H_4 + C_2H_6$	5
N_2 + A	1
CO_2	23
GAS HIGHER HEATING VALUE (Btu/scf)	370
WATER CONTENT (at 100°F)	6%

Because the Purox system uses oxygen instead of air, the product gas is undiluted with atmospheric nitrogen, accounting for its higher energy content per cubic foot. However, the relatively low heating value of the gas compared to natural gas (1000 Btu per cu ft) makes piping long distances uneconomical. In addition, extensive gas scrubbing equipment is required to clean the gas before use.

Figures 5-9 and 5-10 illustrate the Purox process.

Union Carbide built and operated a 200 tpd Purox demonstration facility for 5 years, beginning in 1974, at South Charleston, West Virginia (see Figure 5-11). This plant is no longer in operation, and to date no other plants have been built.

STARVED AIR SYSTEMS
General

Starved air systems, also called modular combustion units, or controlled air incinerators, combine the principles of incineration and pyrolysis. Unprocessed waste is burned in a primary chamber in a reducing atmosphere to minimize particulates in the gas stream. The effluent gas is then burned in a secondary combustion chamber, in an excess air atmosphere, to destroy entrained particulates and unburned hydrocarbons.

Figure 4-32 illustrates the operation of a typical starved air incinerator. Figure 5-12 illustrates the layout of a 100 tpd system in North Little Rock, AK.

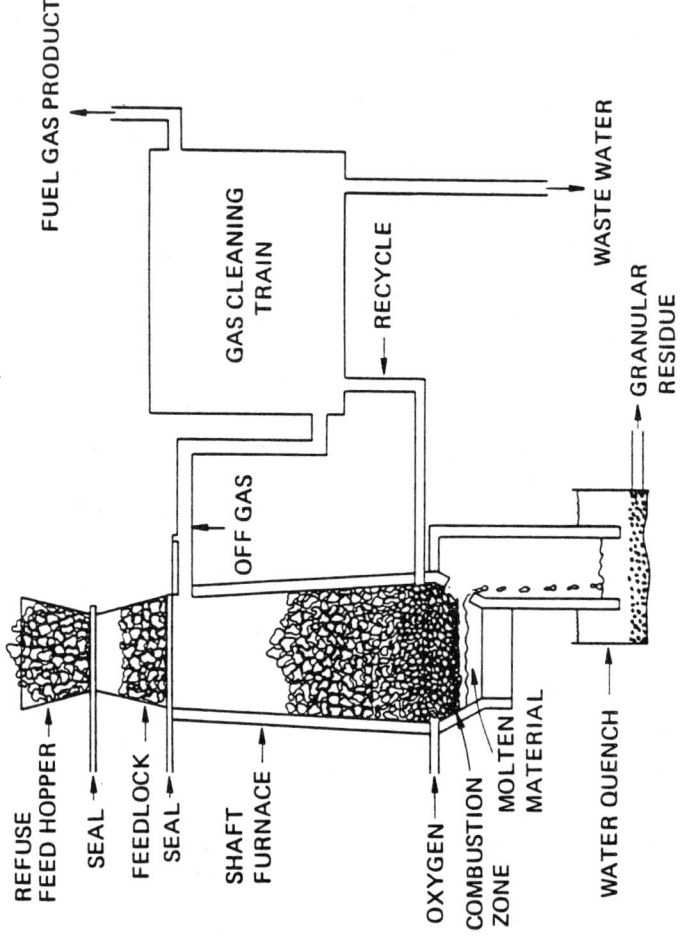

Figure 5-9
Purox Pyrolysis System
(Union Carbide)

Resource Recovery Systems   147

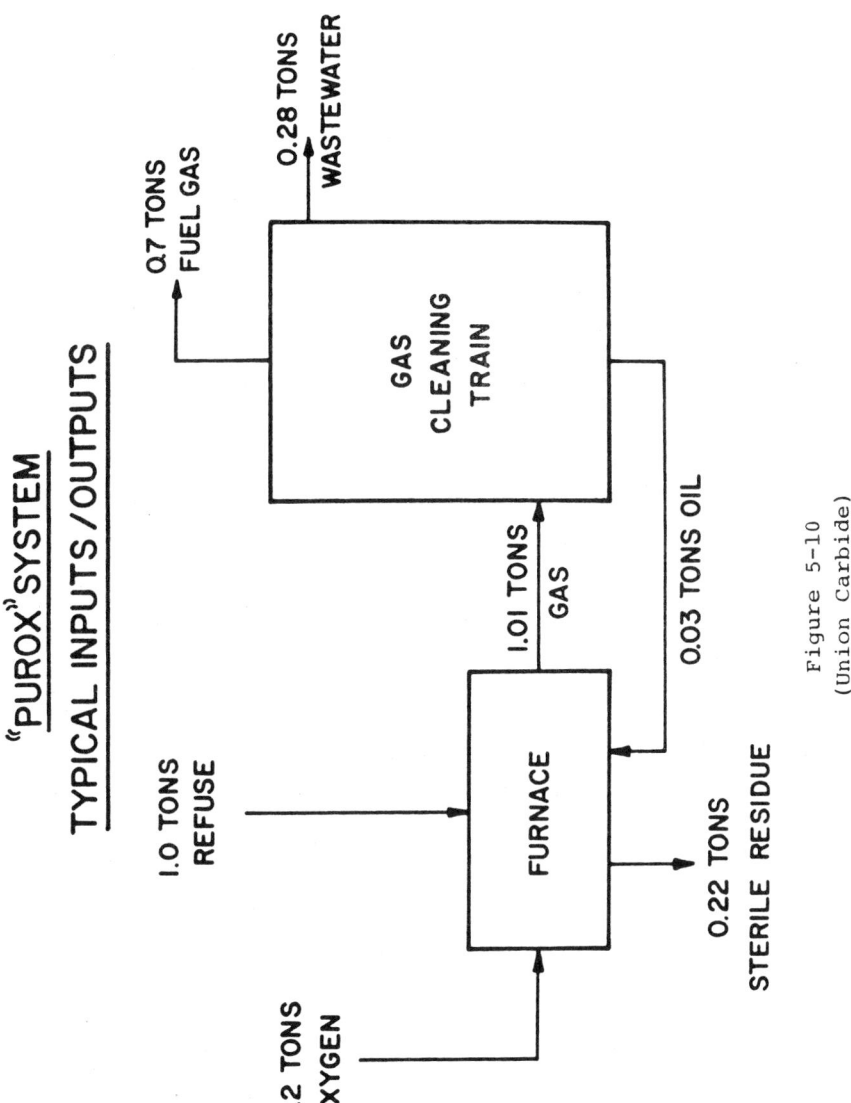

Figure 5-10
(Union Carbide)

Figure 5-11
South Charleston, West Virginia
Purox Facility
(Union Carbide)

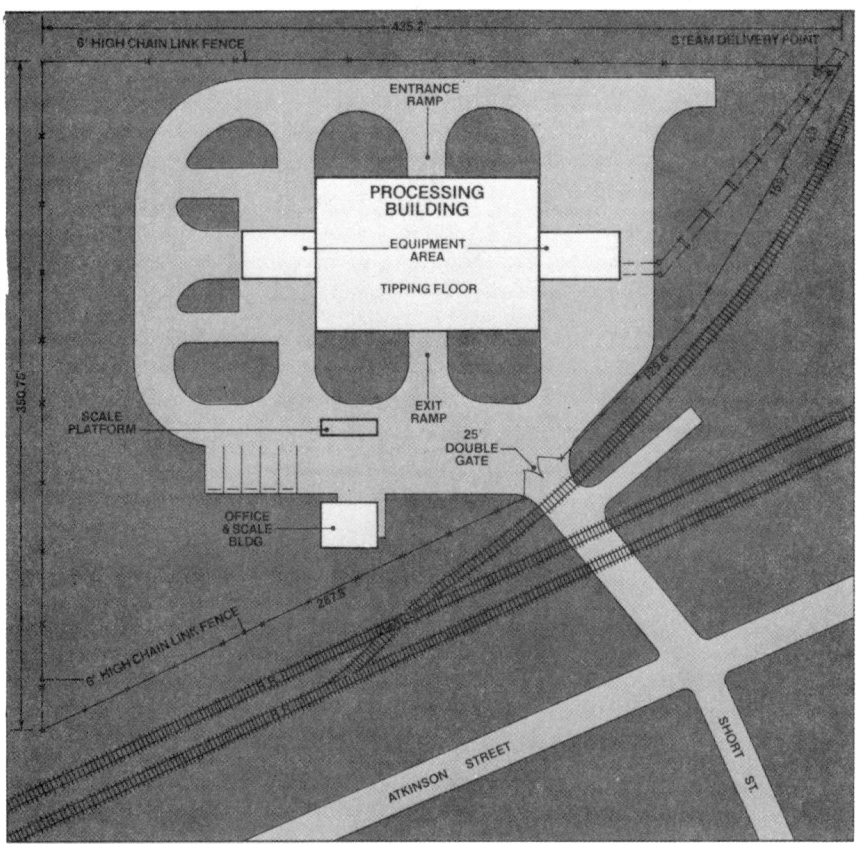

Figure 5-12
North Little Rock Site Plan
(Consumat, Inc.)

Modular combustion systems are available from several manufacturers in the United States today. All share certain distinguishing characteristics:

- o Relatively small capacity compared to field-erected refractory incinerators.

- o Shop fabrication.

- o Use of two-stage, controlled air combustions.

The largest currently operating modular systems utilize modules with a capacity of 25 tons per day each. Fifty ton per day modules are under construction, and 100 ton per day units are under development. The relatively small size of these units makes them suitable for small scale-waste generators including small communities, hospitals, universities and industry. In addition, their small size makes shop fabrication and rail or truck shipment to the job site feasible. Shop fabricated units are significantly less expensive per unit of capacity than similar field erected equipment.

The most important feature of modular units is the use of a two-stage controlled air combustion process. Waste is fed into the first stage, or primary combustion chamber, and burned with less than stoichiometric air. This starved air condition causes most of the volatile fraction to be destroyed pyrolytically, with the required endothermic heat provided by the oxidation of the fixed carbon fraction. The resultant smoke and other pyrolytic products, including primarily volatile hydrocarbons and carbon monoxide, pass to the second stage where additional air is injected to complete combustion. Because combustion reactions and air velocities are maintained at low levels in the primary combustion chamber, particulate entrainment and carryover are minimized.

Although starved air incinerators have been commercially available for over a decade, most early installations were small batch-fed units, without continuous ash removal or heat recovery. Dozens of municipal scale units are in operation today, the oldest beginning operation in 1975. Many of these systems have experienced problems such as:

- poor combustion performance
- failure to meet air pollution standards
- slagging and blockage of underfire air ports
- premature deterioration of refractory

Design changes to overcome these problems have and are being implemented, but it is too early to say if these will solve all problems.

STATUS OF RESOURCE RECOVERY IN THE US

The United States Conference of Mayors tabulates and publishes a semi-annual summary of resource recovery projects in the United States. The latest summary, for March, 1982, is reproduced here as Figure 5-13.

Resource Recovery Activities

The U.S. Conference of Mayors has assumed responsibility for preparation and distribution of this listing of resource recovery facilities in the United States and Canada, formerly published by the National Center for Resource Recovery, Inc.

The report is broken into three segments: (1) facilities that are operating, under construction or nearing construction stages to recover materials and energy from municipal solid waste; (2) projects that recover methane gas from municipal solid waste landfills; and (3) jurisdictions that report being committed to some form of resource recovery, with facilities in various planning stages. The list does not include the growing number of communities that conduct source separation programs and/or magnetically separate ferrous metals from mixed refuse.

Although every effort has been made to ensure that the report is complete and current, the status of many of the projects can change at any time. For clarification or additional information on a specific facility, we suggest that you write directly to the source given for that listing.

The Conference of Mayors is grateful for the contributions and cooperation of each project representative, as well as state and local officials and industry representatives who have helped us to compile this information.

Materials and Energy Recovery Facilities

Location and Major Participants (Who owns; who operates)	Processes	Products (Uses)	Capacity (tons per day)	Capital Costs ($ millions)	Status	Source
ALABAMA						
Huntsville (Redstone Arsenal) U.S. Army; Redstone Arsenal (operator); Sanders & Thomas, Inc. (designer)	Mass burning in modular incinerator	Steam	50	3.2	Under construction; start-up expected in July 1983	Ned Healy U.S. Army Corps of Engineers P.O. Box 2288 Mobile, Ala. 36628 Attn: SAMEN-MM
ARKANSAS						
Batesville City (Mfr: Consumat)	Mass burning in modular incinerator	Steam	50	1.2	Operational	Jim Shirrell, Mayor Municipal Bldg. 170 S. Forth St. Batesville, Ark. 75201
Blytheville City (Mfr: Consumat)	Mass burning in modular incinerator	Steam	70 (to be processed)	N/A	Temporarily closed; awaiting delivery of new incinerators	Tom Little, Mayor City Hall Blytheville, Ark. 72315
North Little Rock City; U.S. Recycle Corp.; Consumat Systems, Inc. (operator) (Mfr: Consumat)	Mass burning in modular incinerator	Steam	100	1.45	Operational	Gene Green Consumat Systems, Inc. P.O. Box 3457 North Little Rock, Ark. 72117
Osceola City (Mfr: Consumat)	Mass burning in modular incinerator	Steam	50	1.1	Operational	R.E. Prewitt, Mayor City Hall Osceola, Ark. 72370
CONNECTICUT						
Bridgeport Conn. Resources Recovery Auth.; Occidental Petroleum Corp. and Combustion Equipment Assoc. (designer/operator); Greater Bridgeport Regional Solid Waste Commission	Shredding, air classification, magnetic separation, Eco-Fuel® II production process	Eco-Fuel® II (powdered fuel) for use in utility boiler, ferrous metals	1800	53	Plant is closed due to CEA's financial difficulties; Occidental has taken over facility and proposed modifications to produce shredded RDF decision expected by June 1982	Lynn C. Healey Executive Assistant Conn. Resources Recovery Authority 179 Allyn St. Hartford, Conn. 06103

Figure 5-13

152 Energy and Resource Recovery from Waste

Location and Major Participants	Processes	Products	Capacity (tons per day)	Capital Costs ($ millions)	Status	Source
Windham Town of Windham (Mfr: Consumat)	Mass burning in modular incinerator	Steam	108	4.125	Operational since Nov. 1981	Louise Guarnaccia First Selectman Town of Windham Town Office, 979 Main St. Willimantic, Conn. 06226
DELAWARE						
Wilmington Delaware Solid Waste Authority; EPA; Raytheon Service Co. (designer)	Shredding, air classification, magnetic separation, froth flotation, other mechanical separation; aerobic digestion	Ferrous and nonferrous metals, glass, RDF, humus	1000 tpd municipal solid waste co-processed with 350 tpd of 20% solids digested sewage sludge	71.3	Under construction; start-up expected in late 1982	Pasquale S. Canzano Chief Engineer Delaware Solid Waste Authority P.O. Box 455 Dover, Del. 19901
FLORIDA						
Dade County (Dade County Solid Waste Resource Recovery Plant) County; Parsons & Whittemore, Inc. (designer/operator)	Hydrasposal™ (wet pulping), magnetic and other mechanical separation	Steam for utility to produce electricity, aluminum, ferrous metals	3000	165	Construction almost complete; startup depends upon resolution of contract dispute between P&W and County	Dennis Carter, Asst. County Manager Room 911 Dade County Courthouse 73 W. Flagler St. Miami, Fla. 33130
Dade County (Agripost Composting Plant) County; Agripost, Inc. (owner/operator/designer)	Composting	Soil conditioner	400	8.5	Contract signed; facility under design (pilot plant)	(Same as previous listing)
Jacksonville (Naval Air Station) U.S. Navy	Shredding, magnetic separation, trommel screening, mass burning in modular incinerator	Steam	40 design; 20 actual	2.8	Operational	R.P. LeVasseur Dir. of Utilities P.O. Box 5, Code 187 Naval Air Station Jacksonville, Fla. 32212
Lakeland City (operator and joint owner with Orlando Utility Commission); C.T. Main, Inc. (power plant designer); Horner & Shifrin, Inc. (waste processing plant designer)	Shredding, magnetic separation, burning RDF with coal	Steam to produce electricity for use by City of Lakeland and Orlando Utility Commission, ferrous metals	300	5 (for waste processing plant)	Under construction; start-up of waste processing plant expected in Spring 1982	Claude Hiers Supt. of Industrial Engin. & Business Affairs Box 368 Lakeland, Fla. 33802
Orange County (Walt Disney World) U.S. Dept. of Energy, Idaho Operations Office; Reedy Creek Utilities Co.; Andco, Inc.	Slagging pyrolysis incineration (Andco-Torrax)	High temperature hot water for heating and cooling	100	15	Under construction; completion expected in June 1982; tests scheduled for approximately 6 months, with municipal waste processing to begin in Oct. 1982 (demonstration plant)	Carl P. Gertz Project Manager U.S. Dept. of Energy 550 Second St. Idaho Falls, Idaho 83401
Pinellas County County; Florida Power Corp.; UOP, Inc.	Mass burning, mechanical separation of metals after burning	Electricity, ferrous and nonferrous metals	2000	160	Construction began Aug. 1980; operational in 3 years	Don F. Acenbrack Director, Solid Waste Mgmt. Dept. Pinellas County 315 Court St. Clearwater, Fla. 33516
Pompano Beach Waste Management, Inc.; U.S. Dept. of Energy; Jacobs Engineering Co. (designer)	Shredding, air classification, magnetic and other mechanical separation, anaerobic digestion of air classified light fraction with sewage sludge	Methane gas, carbon dioxide	50-100	3.65	Operational (demonstration plant)	Ted Sjoberg Dir. of Engineering Waste Management, Inc. 900 Jorie Blvd. Oak Brook, Ill. 60521
HAWAII						
Honolulu City and County of Honolulu; Combustion Engineering, Inc./Amfac, Inc.	Flail milling, magnetic separation, trommel and other screening, secondary shredding, burning RDF in spreader stoker boilers for generation of electricity	Ferrous metals, electricity for utility	1800	110	Negotiations underway to review impact of possible plant site relocation; construction expected to begin in early 1983	Frank Doyle, Chief Div. of Refuse Collection & Disposal Dept. of Public Works City & County of Honolulu Honolulu, Hawaii 96813

Figure 5-13 (continued)

Resource Recovery Systems   153

Location and Major Participants	Processes	Products	Capacity (tons per day)	Capital Costs ($ millions)	Status	Source
IDAHO						
Burley Cassia County (Mfr: Consumat)	Mass burning in modular incinerator	Steam	50	1.5	In shakedown	Norman Dayley County Commissioner Cassia County Court House Burley, Idaho 83318
ILLINOIS						
Chicago (Northwest Waste-to-Energy Facility) City; Metcalf & Eddy, Inc. (designer)	Mass burning in water-wall incinerators, ferrous recovery from ash (intermittent)	Steam for process use on-site and by Brach Candy Co., ferrous metals	1600	23	Operational	Emil Nigro Coordinating Engineer Dept. of Streets & Sanitation Room 700, City Hall Chicago, Ill. 60602
Chicago (Southwest Supplementary Fuel Processing Facility) City; Ralph M. Parsons Co. and Consoer, Townsend & Assoc. (designer)	Shredding, air classification, magnetic separation	RDF for use by utility; ferrous metals	1000	19	Off-stream to review experience and evaluate future operations; decisions pending	(Same as previous listing)
IOWA						
Ames City; Gibbs, Hill, Durham & Richardson, Inc. (designer)	Baling waste paper, shredding, magnetic separation, air classification, screening, other mechanical separation	RDF for use by utility, baled paper, ferrous metals, aluminum, other nonferrous metals	200 (50 tph)	6.3	Operational	Arnold Chantland, Dir. Dept. of Public Works City Hall 5th and Kellog St. Ames, Iowa 50010
KENTUCKY						
Ft. Knox U.S. Army (Mfr: Burnzol)	Mass burning in modular incinerator	Steam	40	N/A	Under construction; start-up expected in mid-1982	J.T. Hutchins Environmental Management Division Chief ATZK-FE Ft. Knox, Ky. 40121
LOUISIANA						
New Orleans City; Waste Management, Inc. (owner/operator); National Center for Resource Recovery, Inc. (designer/implementer)	Shredding, air classification, magnetic and other mechanical separation	Ferrous metals, aluminum, glass	700	9.1	Shredding/landfilling and ferrous recovery operational; aluminum recovery in shakedown; glass recovery discontinued	Clifford Scineaux, Dir. Dept. of Sanitation City Hall New Orleans, La. 70112
MAINE						
Auburn City (Mfr: Consumat)	Mass burning in modular incinerator	Steam	200	3.98	Operational	Robert Belz Public Works Auburn City Hall 45 Spring St. Auburn, Maine 04210
MARYLAND						
Baltimore City; Baltimore County; Northeast Md. Waste Disposal Authority; Wheelabrator-Frye, Inc.	Mass burning in water-wall furnace, electricity generation, ferrous recovery from ash	Electricity for sale to utility; ferrous metals	2000	140	Pyrolysis plant is closed; new plant will be built on same site. Final contract arrangements being completed; revenue bonds to be issued shortly; construction expected in May 1982 with operation in 1985	Michael Gagliardo Northeast Maryland Waste Disposal Authority One E. Redwood St. Suite 300 Baltimore, Md. 21202
Baltimore County County; Maryland Environmental Service; Teledyne National (designer/operator)	Shredding, air classification, magnetic and other mechanical separation, pelletizing	RDF, ferrous metals, glass	1200	8.4	Operational; recovering ferrous metals and producing shredded and pelletized RDF; glass recovery system being developed and refined; aluminum recovery not in use	Kenneth Cramer Teledyne National Padonia Centre, Ste. 401 30 E. Padonia Rd. Timonium, Md. 21093

Figure 5-13 (continued)

Location and Major Participants	Processes	Products	Capacity (tons per day)	Capital Costs ($ millions)	Status	Source
MASSACHUSETTS						
Braintree City (owner/operator); Camp, Dresser & McKee Inc. (designer)	Mass burning in water-wall furnace	Steam (half of steam produced used by Art & Leather Co.)	250	2.8	Operational	Edward Courchene, Supt. Braintree Thermal Waste Reduction Center Ivory Street Braintree, Mass. 02184
East Bridgewater City of Brockton and nearby towns; Combustion Equipment Assoc.; East Bridgewater Assoc.	Shredding; air classification; magnetic separation; other mechanical separation and production of Eco-Fuel® II	Eco-Fuel® II for industrial boiler; ferrous metals	300 tpd being landfilled, with excess trucked to other landfills	10-12	Plant has served as pilot operation for Eco-Fuel® II production and recently for Eco-Fuel briquet production; currently closed due to CEA's financial difficulties; operations continuing as sanitary landfill	M.G. Magoulas Corp. Vice Pres., Engineering Combustion Equipment Assoc. 136 East 57th St. New York, N.Y. 10022
Haverhill & Lawrence Refuse Fuels, Inc. (owner); BE&C Engineers, Inc. (Boeing subsidiary) (design & construction); Cities of Haverhill & Lawrence	Shredding, magnetic separation, trommel screening at recovery facility in Haverhill; burning RDF for cogeneration of steam and electricity in Lawrence	Steam and electricity for industrial use; surplus electricity sold to utility	1300	85	Groundbreaking scheduled for March 1982, with full operation in 3 years	James E. Ricci, Vice Pres. Refuse Fuels, Inc. P.O. Box 187 Bradford, Mass. 01830
North Andover UOP, Inc.; MITRE Corp.; Mass. Bureau of Solid Waste Disposal, Dept. of Environmental Affairs; participating communities	Mass burning in waterwall furnace, electricity generation	Electricity for sale to utility, steam	1500	74	Service agreements being signed with participating communities; groundbreaking expected in 1982	John F. Albis Project Mgr. 128 Main St. North Andover, Mass. 01845
Pittsfield City, Vicon Recovery Assoc.	Mass burning in modular incinerator	Steam	240	6.2+	Operational	Joseph J. Domas, Jr., Pres. Vicon Recovery Assoc. P.O. Box 100, Butler Center Butler, N.J. 07405
Rochester Town and several nearby communities; Energy Answers Corp. (owner); Smith & Mahoney, P.C., and Gordon L. Sutton Assoc. (designers)	Shredding, magnetic separation, burning RDF in semi-suspension stoker-grate boiler, nonferrous recovery from ash, generation of electricity	Electricity for sale to Commonwealth Electric; ferrous and nonferrous metals	1500	120	In design phase; electricity purchase agreement signed; contracts for waste being signed; construction expected to begin in Fall 1982 with operation in 1985	John C. Lemery President Energy Answers Corp. 48 Howard St. Albany, N.Y. 12207
Saugus Thirteen communities including Saugus and part of northern Boston; RESCO (owner/operator)	Mass burning in waterwall furnaces, magnetic separation	Steam for electrical generation and industrial use, ferrous metals	1200	50	Operational	John M. Kehoe, Jr. Wheelabrator-Frye, Inc. Liberty Lane Hampton, N.H. 03842
MICHIGAN						
Detroit City; Combustion Engineering, Inc.	Flail milling, trommel screening, secondary shredding, burning RDF in on-site dedicated boilers, electricity generation in 47 Mw turbo-generator	Steam for Detroit Edison's central heating system; electricity for sale to Detroit Edison; ferrous metals	3000	150	Negotiating with Combustion Engineering prior to contract signing; tax counsel review and permit applications in process; preparing a revenue bond issue with equity participation to finance the facility	Michael Brinker Dept. of Public Works City of Detroit City-County Bldg., Rm. 513 Detroit, Mich. 48226
Genesee Township Township (Mfr: Consumat)	Mass burning in modular incinerator	Steam	100	2	Operation interrupted due to lack of steam customers; negotiating with prospective customer	Hanumanthaiya Marur, P.E. Engineer 1423 E. Bristol Rd. Burton, Mich. 48529
MINNESOTA						
Collegeville St. John's University; Basic Environmental Engineering (designer)	Mass burning in modular incinerator	Steam	65	2.4	In shakedown	Fr. Gordon Tavis St. John's University Collegeville, Minn. 56321

Figure 5-13 (continued)

Resource Recovery Systems

Location and Major Participants	Processes	Products	Capacity (tons per day)	Capital Costs ($ millions)	Status	Source
Duluth Western Lake Superior Sanitary District (owner/operator); Consoer, Townsend & Assoc. (designer)	Shredding, magnetic separation, air classification, secondary shredding, fluidized bed incineration of RDF and sludge	RDF, ferrous metals, steam for heating and cooling of plant and to run process equipment	400 of MSW; 340 of 20% solids sewage sludge	19	In shakedown; full operation expected in June 1982	John Klaers Western Lake Superior Sanitary Dist. 27th Ave. West & The Waterfront Duluth, Minn. 55806
Redwing City (Mfr: Consumat)	Mass burning in modular incinerator	Steam	72	2.5	Under construction; operation scheduled for Nov. 1982	Dean Massett Council Administrator Box 34 Redwing, Minn. 55066
MISSOURI						
Ft. Leonard Wood U.S. Army (Mfr: Environmental Control Products)	Mass burning in modular incinerator	Steam	75	3.3 (approx.)	Under construction; start-up expected in March 1982	Dan Harrison Facility Engineers Building 2200 Ft. Leonard Wood, Mo. 65473
MONTANA						
Livingston Park County; WIDJAC Corp. (Mfr: Consumat)	Mass burning in modular incinerator	Steam	75	3	Under construction; shakedown expected in April 1982	O.M. Wilmot WIDJAC Corp. 10604 N.E. 38th Place Suite 222 Kirkland, Wash. 98033
NEW HAMPSHIRE						
Durham (Lamprey Regional Solid Waste Cooperative) Members of cooperative: 12 towns, 1 city (Mfr: Consumat)	Mass burning in modular incinerator	Steam	108	3.3	Operational	Richard Rugg Administrator Lamprey Regional Solid Waste Cooperative 1 Lamprey Way Durham, N.H. 03824
Groveton City; Groveton Paper Mill, Inc. (Mfr: Environmental Control Products)	Mass burning in modular incinerator	Steam	24	N/A	Operational	Rick Coville Groveton Paper Mill, Inc. Groveton, N.H. 03582
Portsmouth City (Mfr: Consumat)	Mass burning in modular incinerator	Steam	200	6	Under construction; operation expected in July 1982	Calvin Canney City Manager 126 Daniel St. Portsmouth, N.H. 03801
NEW JERSEY						
Ft. Dix U.S. Army; Sanders & Thomas, Inc. (consulting engineer)	Mass burning in modular incinerator	Steam	80	5.5	In final design; construction expected to begin in Summer 1982 with start-up in Fall 1983	Michael Lawrenceau U.S. Army Corps of Engineers New York District 26 Federal Plaza New York, N.Y. 10007
NEW YORK						
Albany City and 10 nearby communities; Smith & Mahoney (designer); Aenco, inc. (processing plant operator); N.Y. State (steam plant owner/operator)	Shredding, magnetic separation, burning RDF in semi-suspension stoker-grate boiler, nonferrous recovery from boiler ash	RDF, ferrous metals, steam for urban heating and cooling, nonferrous metals	750 tons per shift	26.6 (11.6 processing plant; 15 steam plant)	Processing plant operational; steam generating facility in shakedown	Patrick Mahoney Smith & Mahoney 79 N. Pearl St. Albany, N.Y. 12207
Cuba (Cattaraugus County Refuse-to-Energy Facility) Cattaraugus County (owner); Barton & Loguidice, P.C. (designer)	Mass burning in modular incinerator	Steam	108	5.5	Under construction; start-up expected in Oct. 1982	William White Refuse Administrator Cattaraugus County 200 Erie St. Little Valley, N.Y. 14755

Figure 5-13 (continued)

156 Energy and Resource Recovery from Waste

Location and Major Participants	Processes	Products	Capacity (tons per day)	Capital Costs ($ millions)	Status	Source
Glen Cove City (owner/operator); William F. Cosulich and Ernest F.W. Frank (designer)	Mass burning in stoker-fired furnace with centrifuged sewage sludge	Steam for electricity for use at sewage plant and incinerator	250	34 (22 for mass burning unit; 12 for sewage plant)	Sewage plant in operation; incinerator unit under construction; planned start-up July 1982	Joseph P. Hurley Dir. of Public Works City Hall Bridge St. Glen Cove, N.Y. 11542
Hempstead Town; Hempstead Resources Recovery Corp. (subsidiary of Parsons & Whittemore, Inc.) (owner/operator)	Hydrasposal™ (wet pulping), magnetic and other separation, burning of RDF in air-swept spout spreader stoker boilers	Electricity from utility-owned turbine generators, color-sorted glass, aluminum, ferrous metals	2000 (150 tph)	130	Temporarily shut down by joint agreement between Town and HRRC until EPA establishes uniform standards or guidelines for testing of dioxins	Jim McGiffin Acting Plant Manager Hempstead Resources Recovery Corp. P.O. Box 5010 Roosevelt Field Station Garden City East, N.Y. 11530
Monroe County County (owner); Raytheon Service Co. (designer/operator); CH$_2$M Hill (owner's representative)	Shredding, air classification, froth flotation, magnetic and other separation	RDF for use by utility as supplemental boiler fuel, ferrous metals, nonferrous metals, glass	2000	62.2	Recovery facility in shakedown; RDF receiving/storage facility complete; test-burning RDF	Howard Christensen Dir. of Solid Waste Dept. of Public Works 110 Colfax St. Rochester, N.Y. 14606
Niagara Falls Hooker Energy Corp. (Hooker Chemicals & Plastics Corp.) (owner/operator)	Shredding; magnetic separation; burning shredded refuse	Steam and electricity for use by company complex; ferrous metals	2200	100+	Operational at 1000 tpd	James Green Media Relations Hooker Chemical MPO Box 728 Niagara Falls, N.Y. 14302
New York (Betts Avenue Incinerator) City	Mass burning in refractory furnace	Steam for heating and processes in-plant and adjacent City garages	1000 (present throughput 500)	5-waste heat boiler (1965); 24-modifications (1980)	Electrostatic precipitator being installed and other modifications being made; expect to resume 1000 tpd operation upon completion	Paul Gregory Assistant Planner Dept. of Sanitation Office of Resource Recovery 51 Chambers St., Rm. 830 New York, N.Y. 10007
Oceanside Township of Hempstead (owner/operator); Charles R. Velzy (designer)	Mass burning in water-wall furnace	Steam (60,000 lbs./hr.) used in-plant for electricity	750	9	Operational	Al Albanese Supt., Sanitation Township of Hempstead 1600 Merrick Rd. Merrick, N.Y. 11566
Oyster Bay Town of Oyster Bay Industrial Development Agency; Waste Management, Inc. (designer/operator); Lockwood, Kessler & Bartlett, Inc. (consultant)	Mass burning, electricity generation	Electricity	1000	58.5	Contract negotiations under way; construction expected to begin in 1982 with operation in 1985	Karl J. Leupold Chairman Town of Oyster Bay Industrial Development Authority 150 Miller Place Syosset, N.Y. 11791
Westchester County (Peekskill) County & 35 municipalities; Wheelabrator-Frye, Inc.	Mass burning in water-wall furnace, ferrous metal recovery from ash	Steam and electricity for sale to utility, ferrous metals	1890	165	Agreement for N.Y. State portion of funding and agreement for energy sale to Consolidated Edison concluded in Jan. 1982; construction permits received; construction planned for Spring 1982 with operation in Jan. 1985	Edward K. Davies Deputy Commissioner, Solid Waste Mgmt. Rm. 522, County Office Bldg. White Plains, N.Y. 10601

OHIO

Location and Major Participants	Processes	Products	Capacity (tons per day)	Capital Costs ($ millions)	Status	Source
Akron City; Tricil Resources, Inc.	Shredding, magnetic separation, burning RDF in semi-suspension stoker-grate boiler	Steam for urban and industrial heating and cooling, ferrous metals	1000	80	Refuse processing system temporarily shut down for modification of materials handling system; expected to be on-line in Fall 1982; steam being generated by natural gas to supply customers	Dave Chapman 203 Municipal Bldg. 166 South High St. Akron, Ohio 44308

Figure 5-13 (continued)

Resource Recovery Systems   157

Location and Major Participants	Processes	Products	Capacity (tons per day)	Capital Costs ($ millions)	Status	Source
Columbus City; Alden E. Stilson Assoc. (designer)	Shredding, magnetic separation, burning of shredded refuse with supplemental coal in semi-suspension stoker-grate boiler to produce steam and generate electricity	Electricity for city customers	2000 (3000 peak)	154.2	Under construction; operation expected in early 1983	Henry Bell, Supt. Div. of Electricity 50 W. Gay St. Columbus, Ohio 43215
OKLAHOMA						
Miami City; Resource Recovery Systems; Consumat Systems, Inc. (Mfr: Consumat)	Mass burning in modular incinerator	Steam	108	3.14	Under construction; operation expected in Fall 1982	Steve Solomon Resource Recovery Systems 6440 Avondale Dr. Suite 201 Oklahoma City, Okla. 73116
Tulsa Tulsa Authority for the Recovery of Energy; Steam Supply Corp., subsidiary of Alternate Energy Systems, Inc. (owner); Midwesco, Inc. (designer & contractor)	Mass burning, cogeneration of steam and electricity	Steam for sale to Tulsa Refining, Inc.; electricity for sale to Public Service Co. of Okla.	600	36	Contracts in final negotiation; construction expected to begin in late 1982 with operation in mid-1984	Lester M. McCright Alternate Energy Systems, Inc. 4425 East 31st St. Suite J Tulsa, Okla. 74135
OREGON						
Lane County County; Allis-Chalmers Corp. (designer); Western Waste Corp. (operator)	Shredding, air classification, magnetic separation	RDF, ferrous metals	500	2.1	Plant has been accepted by county following cash settlement from Allis-Chalmers; plant is "in mothballs" pending decision to operate; modifications being considered to reduce fuel's ash content	Mike Turner, Adm. Analyst Lane County Solid Waste Mgmt. Div. Environmental Mgmt. Dept. 125 East 8th St. Eugene, Ore. 97401
PENNSYLVANIA						
Harrisburg City; Gannett, Fleming, Corddry and Carpenter, Inc. (designer)	Mass burning of MSW and sewage sludge in waterwall furnace, bulky waste shredding (steam driven), magnetic separation	Steam for utility-owned district heating system and for city-owned sludge drying system, ferrous metals	720	8.3	Operational; sludge drying facility in test	Paul W. Bricker Gannett, Fleming, Corddry and Carpenter, Inc. P.O. Box 1963 Harrisburg, Pa. 17105
RHODE ISLAND						
Warwick R.I. Solid Waste Mgmt. Corp.; Widmer & Ernst	Mass burning, cogeneration of steam and electricity	Steam for state institutional complex, electricity for sale to utility	1200	100	Financing alternatives being explored; construction to begin by mid-1983 with operation in mid-1986; site may be changed to RISWMC Central Landfill in Johnston; decision on site expected by June 1982	Susan J. Baldyga Public Information Coordinator R.I. Solid Waste Mgmt. Corp. 39 Pike St. Providence, R.I. 02903
TENNESSEE						
Crossville City; Environmental Services Corp. (Mfr: Smokatrol; modified by Environmental Control Products, Inc.)	Shredding, mass burning in modular incinerator	Steam	60	1.11	Operational	Nelson C. Walker General Manager Environmental Services Corp. P.O. Box 765 Crossville, Tenn. 38555
Dyersburg City; Colonial Rubber Works, Inc. (Mfr: Consumat)	Mass burning in modular incinerator	Steam	100	2	Operational	Alderman Bob Kirk Colonial Rubber Works, Inc. Dyersburg, Tenn. 38024

Figure 5-13 (continued)

Location and Major Participants	Processes	Products	Capacity (tons per day)	Capital Costs ($ millions)	Status	Source
Gallatin Sumner County; Cities of Gallatin and Hendersonville; Sanders & Thomas, Inc. (designer)	Mass burning in waterwall rotary combustor	Steam for industrial processing and electricity generation	200	9.7	Operational	Jerry H. Metcalf Project Manager P.O. Box 967 Gallatin, Tenn. 37066
Lewisburg City (Mfr: CICO)	Mass burning in modular incinerator	Steam	60	1.75	Operational	John D. Lambert City Manager 505 Ellington Pkwy. Route 1 Lewisburg, Tenn. 37091
Nashville Nashville Thermal Transfer Corp; I.C. Thomasson & Assoc., Inc. (designer)	Mass burning in waterwall incinerator	Steam for urban heating and cooling	530 (processing 400)	24.5	Operational	Milton E. Kirkpatrick Exec. V.P. & Gen. Mgr. Nashville Thermal Transfer Corp. 110 First Ave. South Nashville, Tenn. 37201

TEXAS

Location and Major Participants	Processes	Products	Capacity (tons per day)	Capital Costs ($ millions)	Status	Source
Gatesville (Texas Dept. of Corrections) Texas Dept. of Corrections (Mfr: Consumat)	Mass burning in modular incinerator	Steam	7	.2	Operational	R.E. Howell Chief, Bldg. & Eng. Mgmt. Construction Div. Texas Dept. of Corrections P.O. Box 99 Huntsville, Texas 77340
Palestine (Beto Unit, Texas Dept. of Corrections) Texas Dept. of Corrections (Mfr: Consumat)	Mass burning in modular incinerator	Steam	28	3	Operational	(same as Gatesville, Texas)

VERMONT

Location and Major Participants	Processes	Products	Capacity (tons per day)	Capital Costs ($ millions)	Status	Source
Burlington City; University of Vermont; Medical Center Hospital of Vermont; William F. Cosulich Assoc. (consulting engineer)	Mass burning, ferrous recovery from ash	Steam for use in district heating loop, ferrous metal	120	11.5	In final design; construction expected to begin in July 1982 with operation in April 1984	James R. Ogden Supt. of Streets P.O. Box 849 Burlington, Vt. 05402

VIRGINIA

Location and Major Participants	Processes	Products	Capacity (tons per day)	Capital Costs ($ millions)	Status	Source
Hampton City; NASA Langley Research Center; U.S. Air Force at Langley Field; J.M. Kenith Co. (designer/builder)	Mass burning in waterwall furnace	Steam for use by NASA Langley Research Center	200	10.3	Operational	Frank H. Miller, Jr. Dir. of Public Works Hampton, Va. 23669
Harrisonburg City (owner & operator); William F. Cosulich Assoc. (consulting engineer)	Mass burning	Steam	100	8	Under construction; operation expected in Nov. 1982	John E. Driver Asst. City Manager Municipal Bldg. 345 S. Main St. Harrisonburg, Va. 22801
Newport News (Ft. Eustis) U.S. Army (Mfr: Consumat)	Mass burning in modular incinerator	Steam	40	1.4	Operational	Chris Wenk Area Engineer Southern Va. Area Office P.O. Drawer B Ft. Eustis, Va. 23604
Norfolk (Norfolk Naval Station) U.S. Navy (owner); Public Works Center, Norfolk Naval Station (operator)	Mass burning in waterwall furnace	Steam for use by facilities at Norfolk Naval Station	360 (two 180-tpd boilers operated alternately)	2.2 (1967)	Operational	Commanding Officer Navy Public Works Center Attn: Director of Utilities Norfolk, Va. 23511

Figure 5-13 (continued)

Location and Major Participants	Processes	Products	Capacity (tons per day)	Capital Costs ($ millions)	Status	Source
Petersburg United Bio-Fuel Industries, Inc.; Teledyne National; Raphael Katzen Assoc.; Foster Wheeler Synfuels Corp.	Phase I—shredding, magnetic and other separation, burning of RDF for electricity generation; Phase II—ethanol production using licensed process of enzymatic hydrolysis of cellulose to alcohol	Phase I—ferrous and nonferrous metals, glass, electricity for sale to utility, steam for in-plant use; Phase II—ethanol, CO_2, dried grain supplement (DGS), distiller's dried grain supplement (DDGS)	3000 (peak)	88 (Phase I) 135 (Phase II)	Preliminary design completed; groundbreaking expected in June 1982 with start-up in early 1984 for Phase I	Francis B. Richerson Director of Engineering United Bio-Fuel Industries, Inc. P.O. Box 1312 Richmond, Va. 23210
Portsmouth (Norfolk Naval Shipyard) U.S. Navy (owner); Public Works Dept., Norfolk Naval Shipyard (operator)	Mass burning in water-wall furnace	Steam for use by facilities at Naval Shipyard	160 (two 80-tpd boilers, operated alternately)	4.5	Operational	Commander Norfolk Naval Shipyard Attn: Public Works Officer Portsmouth, Va. 23709
Portsmouth (Southeastern Tidewater Energy Project) Southeastern Public Service Authority of Va.; Henningson, Durham & Richardson (architect/engineer); Day & Zimmerman (construction manager); Norfolk Naval Shipyard	Shredding, air classification, magnetic and other separation	RDF for burning in power plant at Naval Shipyard, ferrous and nonferrous metals	2000	70	Concept has changed to RDF production instead of steam/electricity generation; contracts in approval process; operation projected for late 1986	Durwood S. Curling Executive Director Southeastern Tidewater Energy Project 16 Koger Executive Center, Suite 129 Norfolk, Va. 23502
Salem City (Mfr: Consumat)	Mass burning in modular incinerator	Steam	100	1.9	Operational	William Paxton, Jr. City Manager P.O. Box 869 Salem, Va. 24153

WASHINGTON

Location and Major Participants	Processes	Products	Capacity (tons per day)	Capital Costs ($ millions)	Status	Source
Tacoma City (owner/operator); Boeing Engineering (designer)	Shredding, air classification, magnetic separation	RDF, ferrous metals	500	2.5	Operational; running periodically to produce RDF for test burning	Bill Larson, Proj. Mgr. Refuse Utility 740 St. Helens Ave. Rm. 304 Tacoma, Wash. 98402

WISCONSIN

Location and Major Participants	Processes	Products	Capacity (tons per day)	Capital Costs ($ millions)	Status	Source
Madison City and M.L. Smith Environmental (designer); Madison Gas & Electric Co. (RDF user)	Shredding, magnetic separation, trommel screening, secondary shredding	RDF burned by utility for electricity generation; ferrous metals	400 (max) (250 being processed)	2.5	Operational	Robert Vetter Div. of Engineering Rm. 115, City-County Bldg. Madison, Wis. 53709
Milwaukee City; Americology Div. of American Can Co. (owner/operator); Bechtel, Inc. (designer)	Shredding, air classification, Americology separation (nonferrous and glassy aggregate separation not in use)	RDF for use by utility, bundled paper and corrugated, ferrous metals	1600	18 (plus 4.2 at Wis. Electric Power Co.)	Reached design capacity in April 1980; maintained until Sept. 1980; temporarily shut down pending negotiations with the City of Milwaukee and WEPCO regarding boiler slagging problems on coal only, which are aggravated when co-firing RDF	George Mallan Dir. of Operations & Marketing Americology American Can Co. GOP #8 Greenwich, Conn. 06830
Waukesha City; Donohue & Assoc. (incinerator designer); Sanders & Thomas, Inc. (heat recovery system designer)	Mass burning in refractory furnace	Steam for local industry and sewage treatment plant	175 (design) (120 burning)	Incinerator 1.7 (1971) Heat recovery system 3.9 (1979)	Incinerator operating since 1971; waste heat recovery boiler added in 1979; operating and sending steam to local industry and sewage plant	Rodney Vanden Noven Dir. of Public Works 201 Delafield St. Waukesha, Wis. 53186

CANADA

ONTARIO

Location and Major Participants	Processes	Products	Capacity (tons per day)	Capital Costs ($ millions)	Status	Source
Hamilton Regional Municipality of Hamilton-Wentworth (owner); Tricil Ltd. (operator)	Shredding, magnetic separation, burning in dedicated spreader stoker boiler	Ferrous metal, steam	500	9*(1972)	Operational since 1972; 3.2 MW turbine generator is being added and expected to be operating in July 1982	Joseph Kennedy Director, Energy Recovery Projects Tricil Ltd. 101 Queensway West, Suite 400 Mississauga, Ontario L5B 2P7

Figure 5-13 (continued)

Location and Major Participants	Processes	Products	Capacity (tons per day)	Capital Costs ($ millions)	Status	Source
Toronto Ontario Ministry of the Environment; Browning-Ferris Industries (operator)	Shredding, air classification, secondary shredding, screening, mass burning in modular incinerator with heat recovery, ferrous cleaning; also transfer operation	Ferrous metal, RDF, compost; hot water for plant heating	Resource recovery—200; transfer facility—600	15*	Operational	Neal R. Ahlberg Plant Manager Ontario Centre for Resourcce Recovery 35 Vanley Crescent Downsview, Ontario M3J 2B7
PRINCE EDWARD ISLAND						
Parkdale Prince Edward Island Energy Corp. (owner); Tricil Ltd. (full-service contractor)	Mass burning in modular incinerator	Steam	108	8.2*	Start-up expected in Dec. 1982	(Same as Hamilton, Ontario)

*Canadian dollars.

Methane Recovery from Landfills

Location and Major Participants	Output or Gas Produced; Million ft³/day	Capital Costs ($ millions)	Status	Source
CALIFORNIA				
Azusa Azusa Land Reclamation Co. (wholly owned subsidiary of the Southwestern Portland Cement Co.)	Low Btu gas	N/A	Operational	F.T. Sheets III Azusa Land Reclamation Co. 1201 W. Gladstone St. Azusa, Calif. 91702
Brea (Olinda Landfill) Getty Synthetic Fuels, Inc.; Orange County	Gas to power generator for electricity production	N/A	Operation scheduled for late 1982	Freederick C. Rice Getty Synthetic Fuels, Inc. 2750 Signal Parkway Signal Hill, Calif. 90806
Carson Watson Biogas Systems; SCS Engineers, Inc.	Medium Btu gas to power generators, producing electricity for sale to utility (1.7 Mw)	N/A	Collection system complete; operations expected in Feb. 1983	Joseph V. Seruto, Pres. Watson Biogas Systems 22010 S. Wilmington Ave. Suite 207 Carson, Calif. 90745
Corona Watson Biogas Systems; Lockman and Assoc.	Medium Btu gas to power generators, producing electricity for sale to utility (5 Mw)	N/A	Contracts signed with City and Southern Calif. Edison; operation expected in June 1983	(Same as Carson, Calif.)
City of Industry (Industry Hills Convention Center) City; SCS Engineers, Inc.; National Engineering Co.	Medium Btu gas; .5 (approx.)	.45	Operational	Robert Stearns SCS Engineers, Inc. 4014 Long Beach Blvd. Long Beach, Calif. 90807
Duarte Watson Biogas Systems; Lockman and Assoc.	Medium Btu gas to power generators, producing electricity for sale to utility (2.3 Mw)	N/A	Collection system under construction; operations expected in June 1982	(Same as Carson, Calif.)
Los Angeles (Bradley East Landfill) Genstar Gas Recovery Systems, Inc.	Medium Btu gas; 2.4	N/A	Operational	Kenneth Wuest Genstar Gas Recovery Systems, Inc. 177 Bovet Rd., Suite 550 San Mateo, Calif. 94420
Martinez Getty Synthetic Fuels, Inc.; Acme Fill Corp.; Contra Costa County Sanitation District	Medium Btu gas; 2.0	N/A	Operational since Feb. 1982	(Same as Brea, Calif.)
Monterey Park Getty Synthetic Fuels, Inc.; Operating Industries, Inc.; Southern California Gas Co.	High Btu gas; 4.0	N/A	Operational	(Same as Brea, Calif.)

Figure 5-13 (continued)

Resource Recovery Systems 161

Location and Major Participants	Output or Gas Produced; Million ft³/day	Capital Costs ($ millions)	Status	Source
Mountain View City of Mountain View; EPA; Pacific Gas & Electric Co.; Dept. of Energy	High Btu gas; 0.5	.85	Demonstration plant; currently operating and producing 0.3 MMSCFD of treated gas with a HHV of 850-950 Btu/SCF; expansion currently under investigation	Max Blanchet Pacific Gas & Electric Co 245 Market St. San Francisco, Calif. 94106
Palos Verdes Getty Synthetic Fuels, Inc.; Los Angeles County Sanitation Dist.; Southern California Gas Co.	High Btu Gas; .75	N/A	Operational	(Same as Brea, Calif.)
San Fernando Getty Synthetic Fuels, Inc.; Browning-Ferris Industries; Newhall Refinery	Medium Btu gas; 1.0	N/A	Operational since Nov. 1981	(Same as Brea, Calif.)
San Leandro Getty Synthetic Fuels, Inc.; Oakland Scavenger Co.; Domtar Gypsum America	Medium Btu gas; 3.0	N/A	Operational	(Same as Brea, Calif.)
Sun Valley (Sheldon-Arleta Landfill Gas Recovery Project) City of Los Angeles Departments of Public Works and Water & Power	Low Btu gas; 2.8	2.5	Compressor station undergoing equipment modifications; operations expected to resume in Fall 1982	Mike Miller Sanitary Engineer L.A. Bureau of Sanitation Room 1410, City Hall East Los Angeles, Calif. 90012
Wilmington Watson Biogas Systems; SCS Engineers, Inc.	2.5	N/A	Operational	(Same as Carson, Calif.)

COLORADO

Location and Major Participants	Output or Gas Produced; Million ft³/day	Capital Costs ($ millions)	Status	Source
Denver City of Denver; Adams County; Watson Biogas Systems; SCS Engineers, Inc.	Medium Btu gas for sale to industrial user; 1.5	N/A	Construction to begin in June 1982; operation expected in April 1983	(Same as Carson, Calif.)

ILLINOIS

Location and Major Participants	Output or Gas Produced; Million ft³/day	Capital Costs ($ millions)	Status	Source
Calumet City Getty Synthetic Fuels, Inc.; Waste Management, Inc.; Natural Gas Pipeline Co. of America	High Btu gas; 2.5	N/A	Operational	(Same as Brea, Calif.)

MICHIGAN

Location and Major Participants	Output or Gas Produced; Million ft³/day	Capital Costs ($ millions)	Status	Source
Riverview Watson Biogas Systems; SCS Engineers	Medium Btu gas for sale to industrial user; 2.5	N/A	Contracts with city signed; applications for construction permits submitted; user negotiations proceeding	(Same as Carson, Calif.)

NEW JERSEY

Location and Major Participants	Output or Gas Produced; Million ft³/day	Capital Costs ($ millions)	Status	Source
Cinnaminson Sanitary Landfill, Inc.; Public Service Electric & Gas Co.; Hoeganaes Corp.	Medium Btu gas (570 Btu/SCF); 1 (Used in-plant by Hoeganaes Corp.)	N/A	Operational; modifications planned to improve service reliability and add other customers	Tom Sharp Public Service Electric & Gas Co. of N.J. 80 Park Plaza T-16A Newark, N.J. 07101

NEW YORK

Location and Major Participants	Output or Gas Produced; Million ft³/day	Capital Costs ($ millions)	Status	Source
Staten Island (Fresh Kills Landfill) Brooklyn Union Gas Co., Inc.; New York City Office of Resource Recovery & Waste Disposal; N.Y. State Energy Research & Development Authority; U.S. Dept. of Energy; Leonard S. Wegman, Inc.	0.05	.33	RD&D program to use raw landfill gas in an internal combustion engine to produce electricity concluded as of Sept. 1981; final report to be issued by N.Y. State Energy Research & Development Authority	Anthony Giuliani Brooklyn Union Gas Co., Inc. 195 Montague St. Brooklyn, N.Y. 11202
Staten Island (Fresh Kills Landfill) Getty Synthetic Fuels, Inc.; City of New York; Brooklyn Union Gas Co.	High Btu gas; 5.0	N/A	Operations scheduled to begin in April 1982	(Sames a Brea, Calif.)

Figure 5-13
(continued)

Location and Major Participants	Output or Gas Produced; Million ft³/day	Capital Costs ($ millions)	Status	Source
NORTH CAROLINA				
Winston-Salem City	Medium Btu gas (burned to generate electricity to power sewage treatment plant)	Less than $25,000 for wells and pipeline	Operational	Lee Byerly Supervisor Archie Elledge Wastewater Treatment Plant 2801 Griffith Rd. Winston-Salem, N.C. 27103

Planned Resource Recovery Facilities

ALABAMA
Opelika

CALIFORNIA
Alameda
Berkeley
Eureka
Gardena
Long Beach
Los Angeles
North Santa Clara County
Point Richmond
San Diego
San Francisco
Ukiah

CONNECTICUT
Hartford
Naugatuck
New Haven
North Haven
Wallingford

FLORIDA
Broward County
Hillsborough County

GEORGIA
Savannah

IDAHO
Bannock County

MASSACHUSETTS
Plainville (128 West)
Springfield

MINNESOTA
St. Paul/Ramsey County

MISSOURI
Kansas City
St. Louis
Springfield

MONTANA
Laurel

NEW JERSEY
Camden County
East Brunswick
Essex County
Ocean County
Union County

NEW YORK
Babylon, Huntington (Multi-Town Authority)
Broome County
Dutchess County
New York
North Hempstead
Oneida County
Onondaga County
Washington County

NORTH DAKOTA
Williston

OHIO
Cincinnati
Cuyahoga County

OREGON
Marion County
Portland

TENNESSEE
Memphis
Nashville

UTAH
Salt Lake City

VIRGINIA
Richmond

WISCONSIN
Appleton
Eau Claire
Green Bay
North Central Wisconsin

Figure 5-13
(continued)

RECOVERY EFFICIENCY

One important criteria for comparison between competing resource recovery technologies is efficiency in energy recovery. Since several processing systems are themselves substantial energy consumers, it is important to compare systems based on net, not gross, efficiency, subtracting processing energy from output.

Table 5-1 presents a comparison of selected resource recovery systems energy recovery efficiency.

COSTS

Figures 5-14, through 5-27 present plots of cost versus size for selected resource recovery facilities.[4] All costs are in June 1981 dollars; capital costs are plotted as a function of installed capacity, O&M costs as a function of average daily throughput. Confidence interval limits are drawn to include the range of costs that are applicable 95% of the time. These confidence intervals plus the equations noted in each of these figures were developed using regression analysis techniques which are described, with plants that were used in this analysis, in Reference 4.

TABLE 5-1

COMPARISON OF ENERGY RECOVERY EFFICIENCIES, PERCENT OF GROSS BTU

Process	Gross Output	Consumed in Processing	Net Output	Boiler Efficiency	Available as Steam	Lbs Steam/Lb MSW [1]
Mass-burning	-	-	-	60-70	60-70	2.11-2.46
Fluff RDF	78	8	70	70	49	1.72
Dust RDF	78-84	4	74-80	78	58-62	2.04-2.18
Coarse RDF	90-95	3-7	83-92	75	62-69	2.18-2.43
Wet Pulp RDF	82	6	76	65	49	1.72
Pyrolysis (Purox)	74	10	64	90	58	2.04
Pyrolysis (Torrax)	101	17	84	69	58	2.04

[1] Assumes 4500 Btu/lb in MSW; steam at 750°F, 600 psig, 1150 Btu/lb, 10% feedwater losses.
Then: lbs steam/lb MSW = 4500 × % available as steam × 0.9/1150

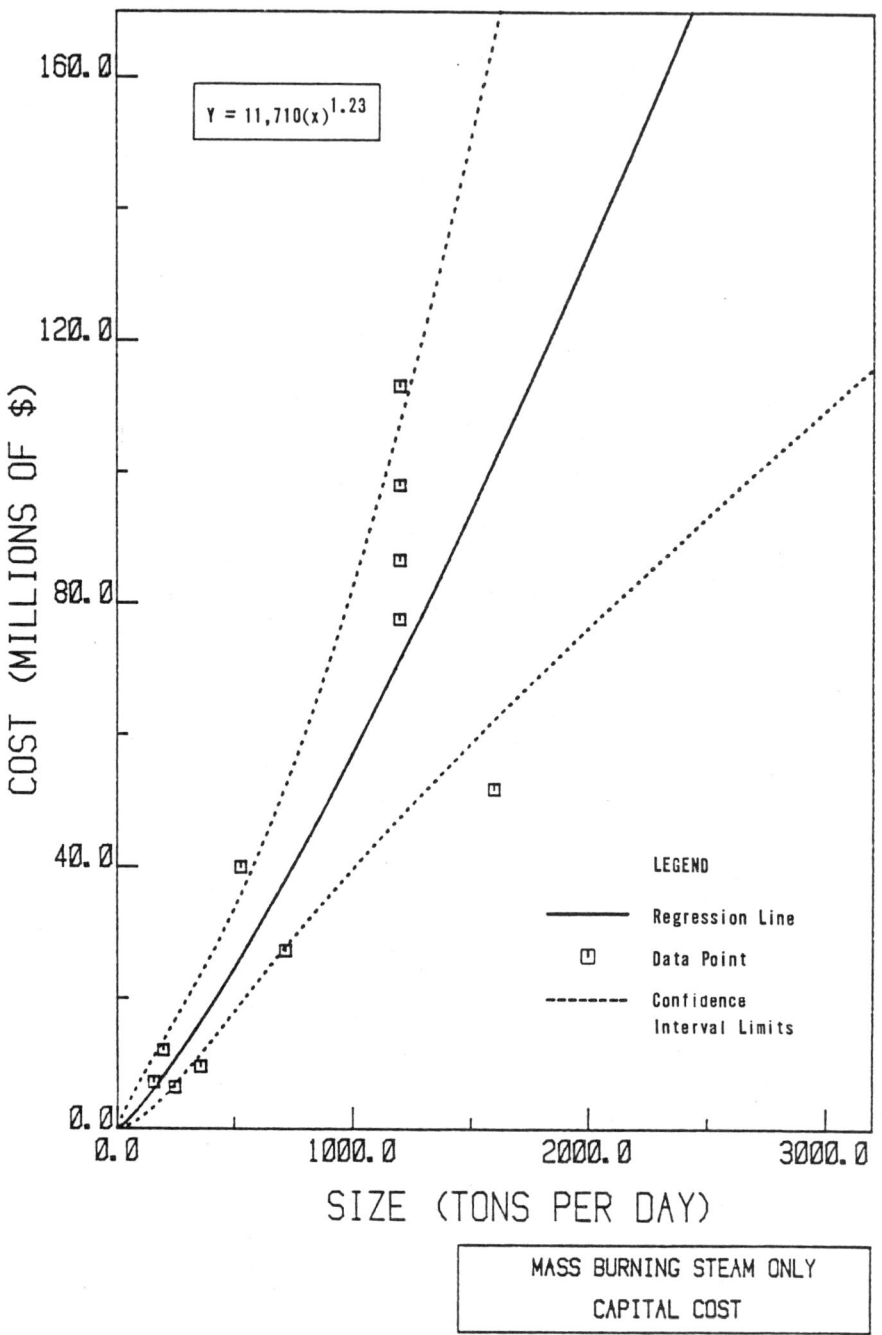

Figure 5-14

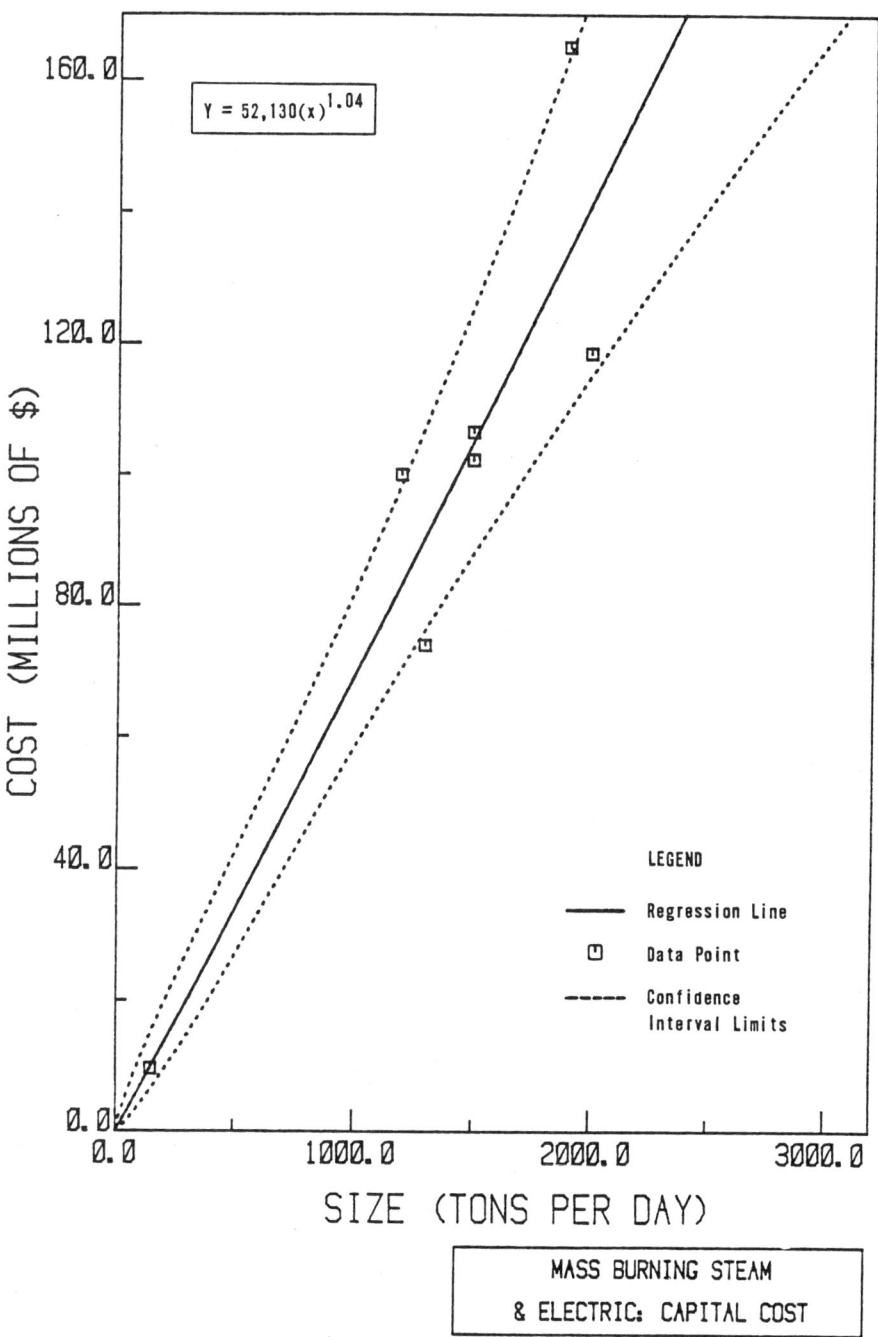

Figure 5-15

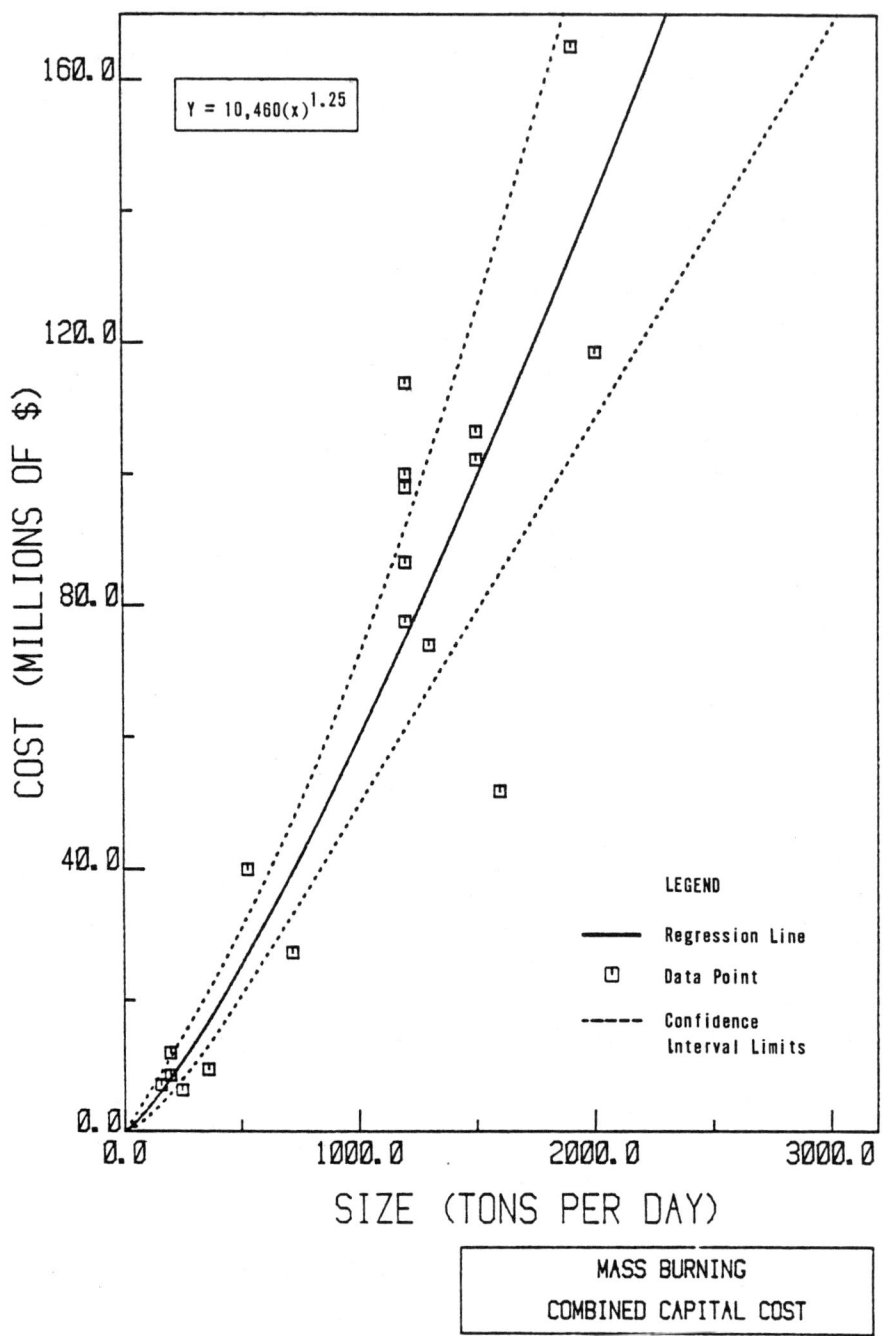

Figure 5-16

Resource Recovery Systems 167

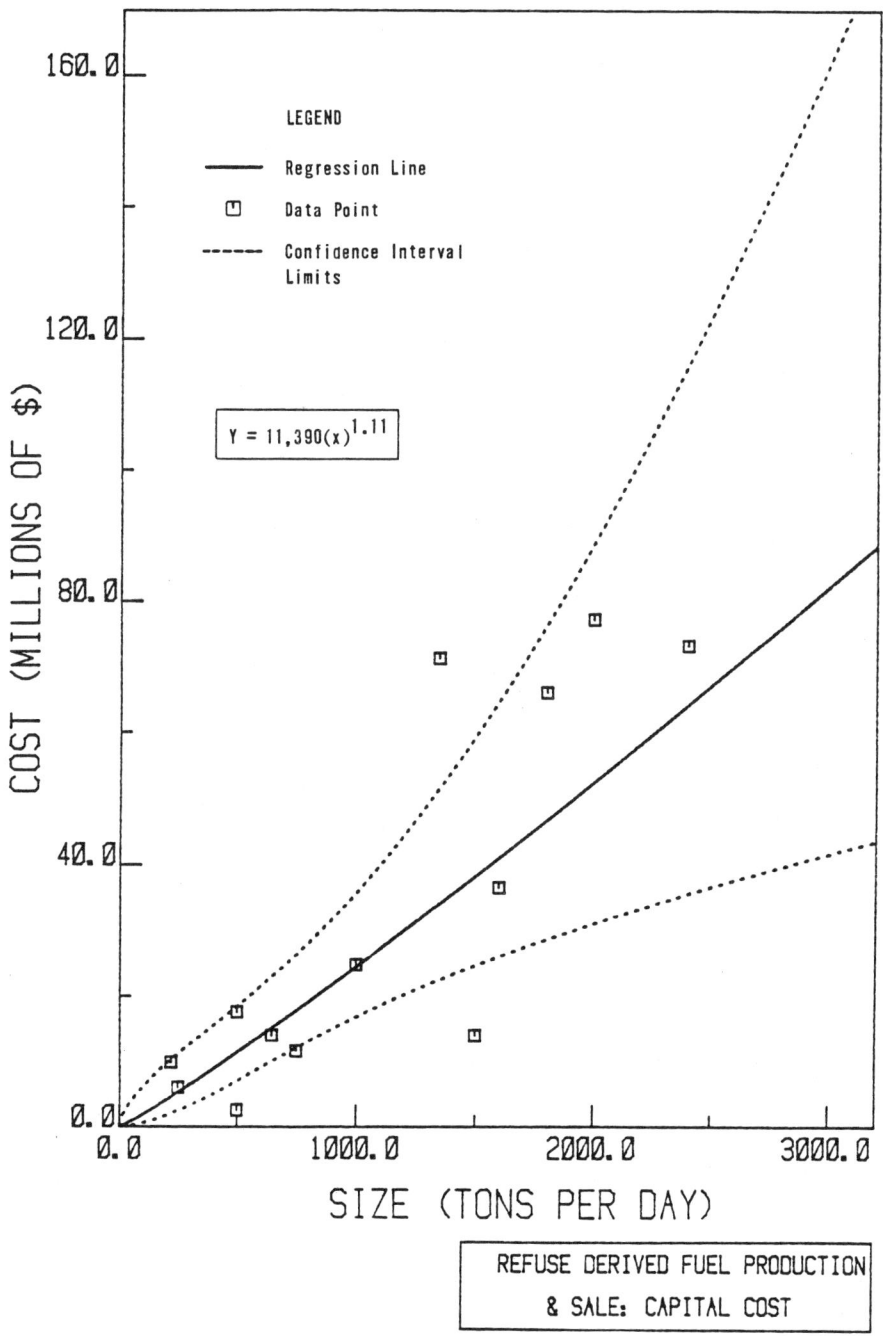

Figure 5-17

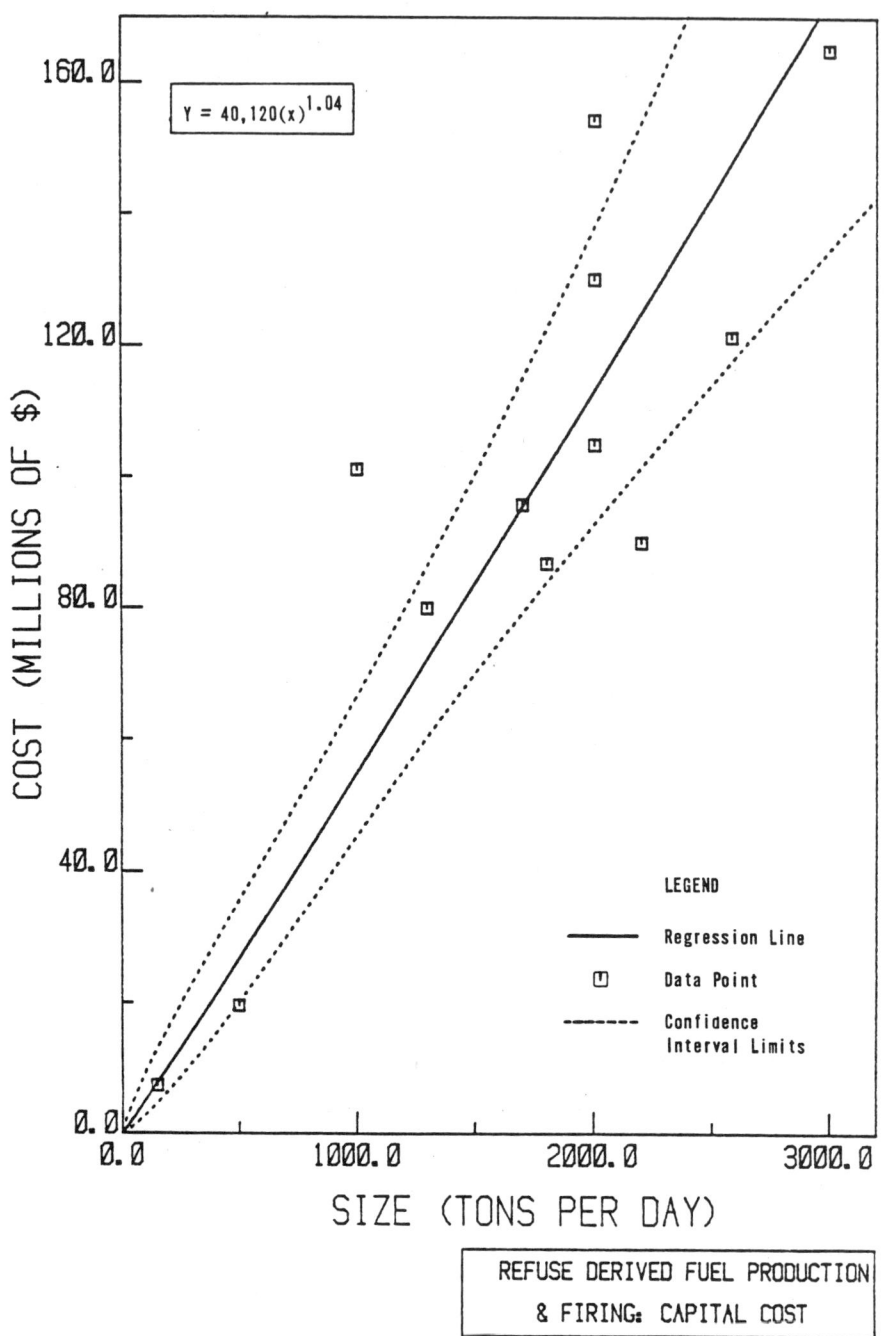

Figure 5-18

Resource Recovery Systems 169

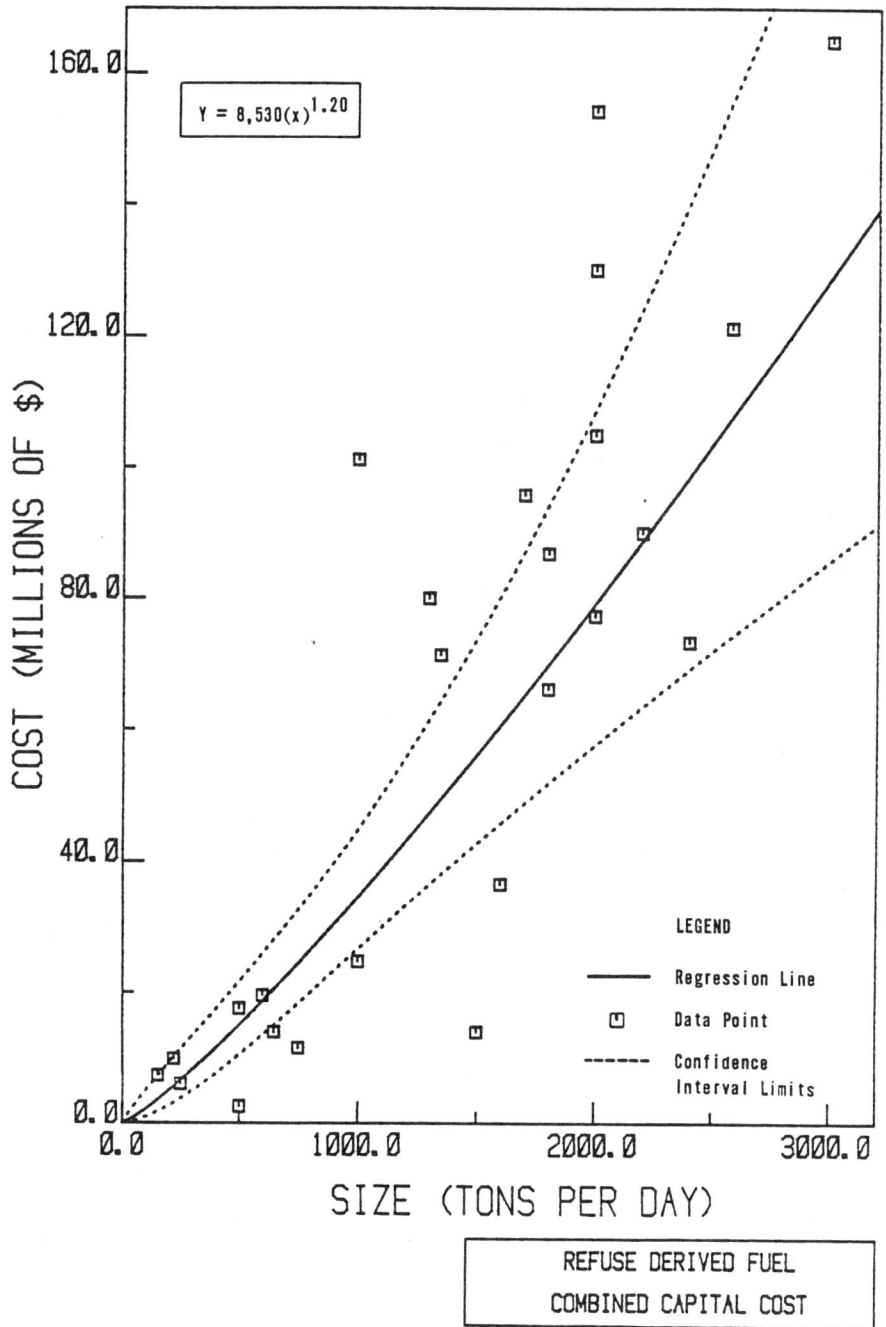

Figure 5-19

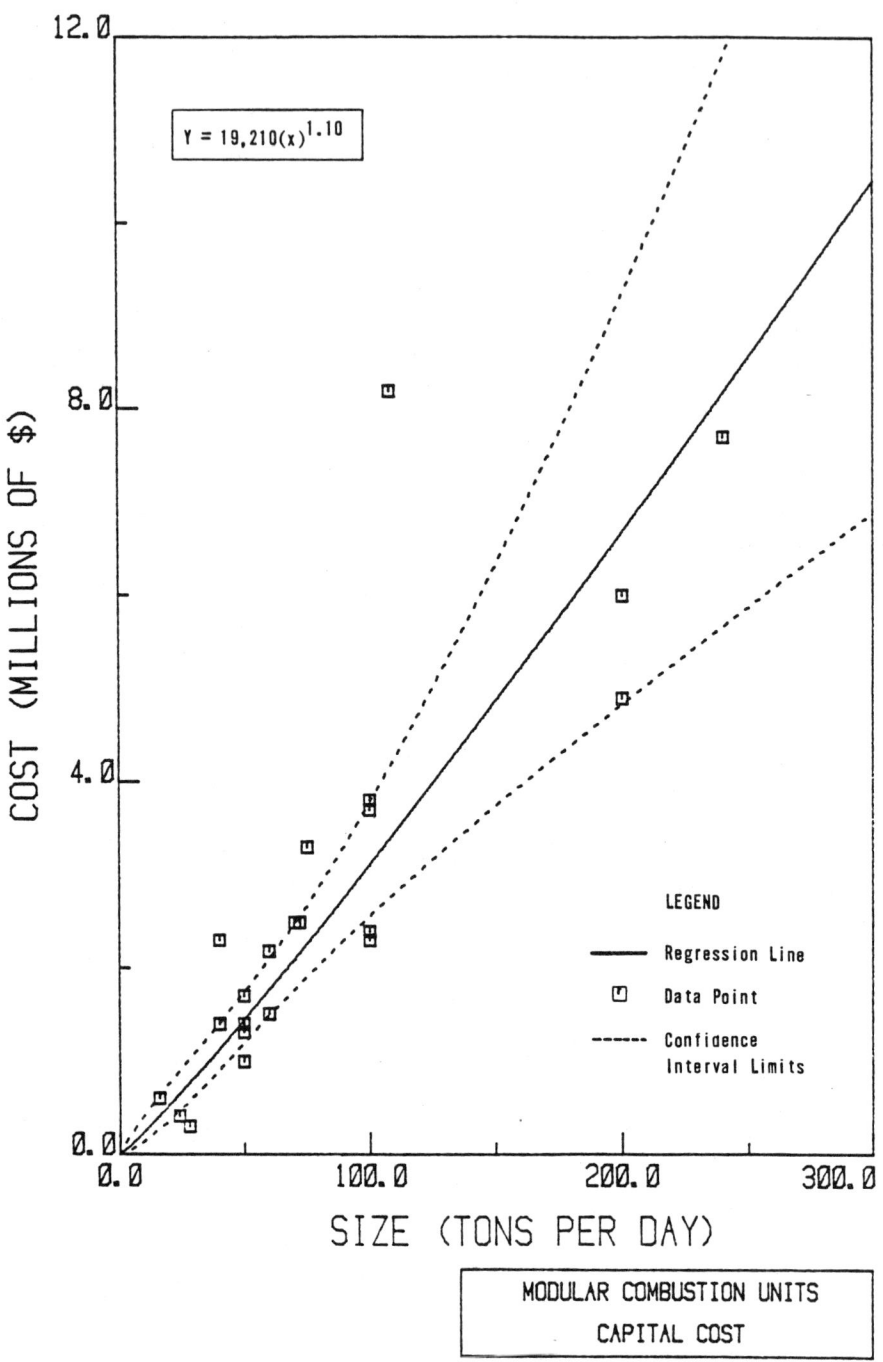

Figure 5-20

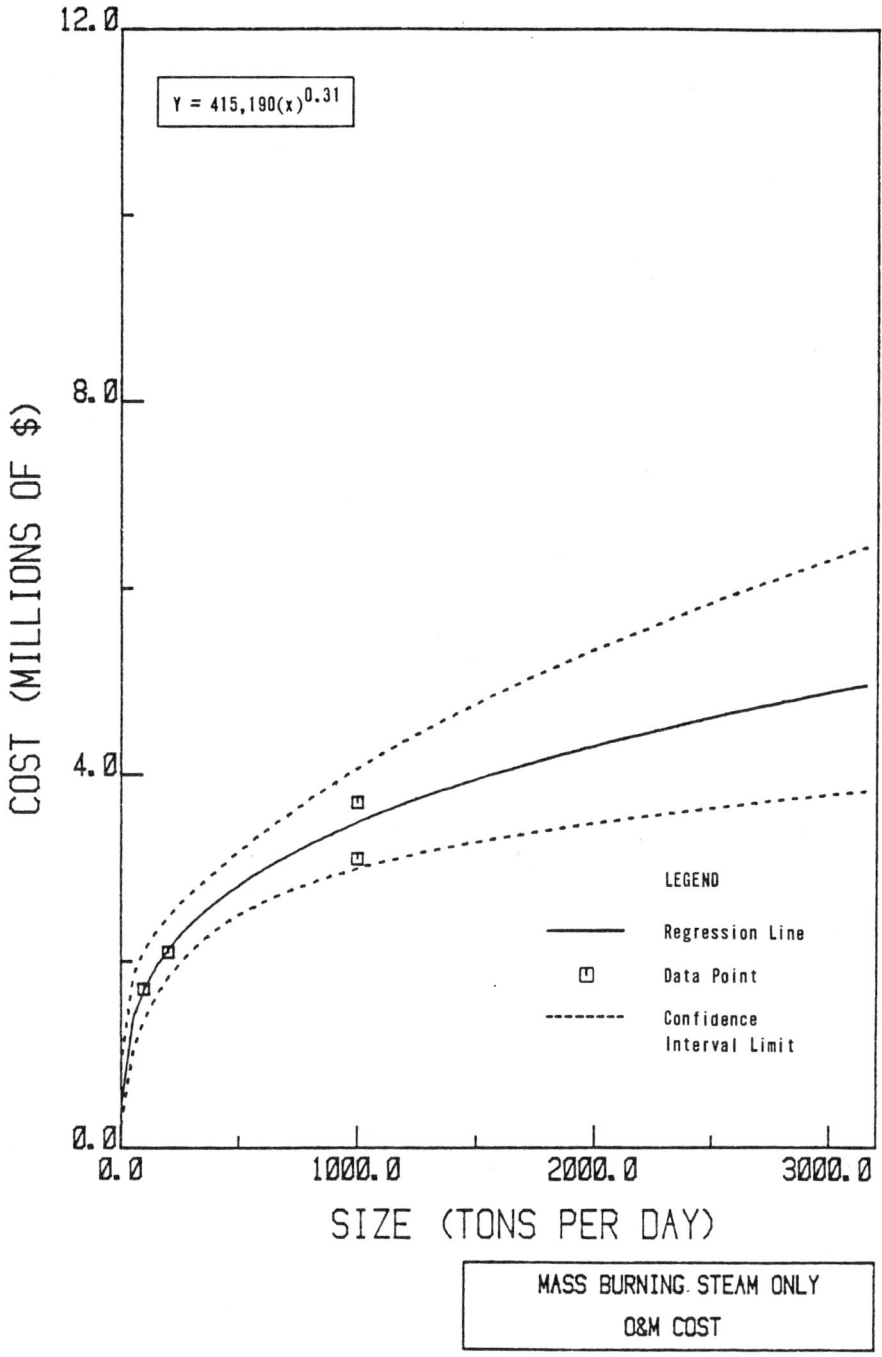

Figure 5-21

172   Energy and Resource Recovery from Waste

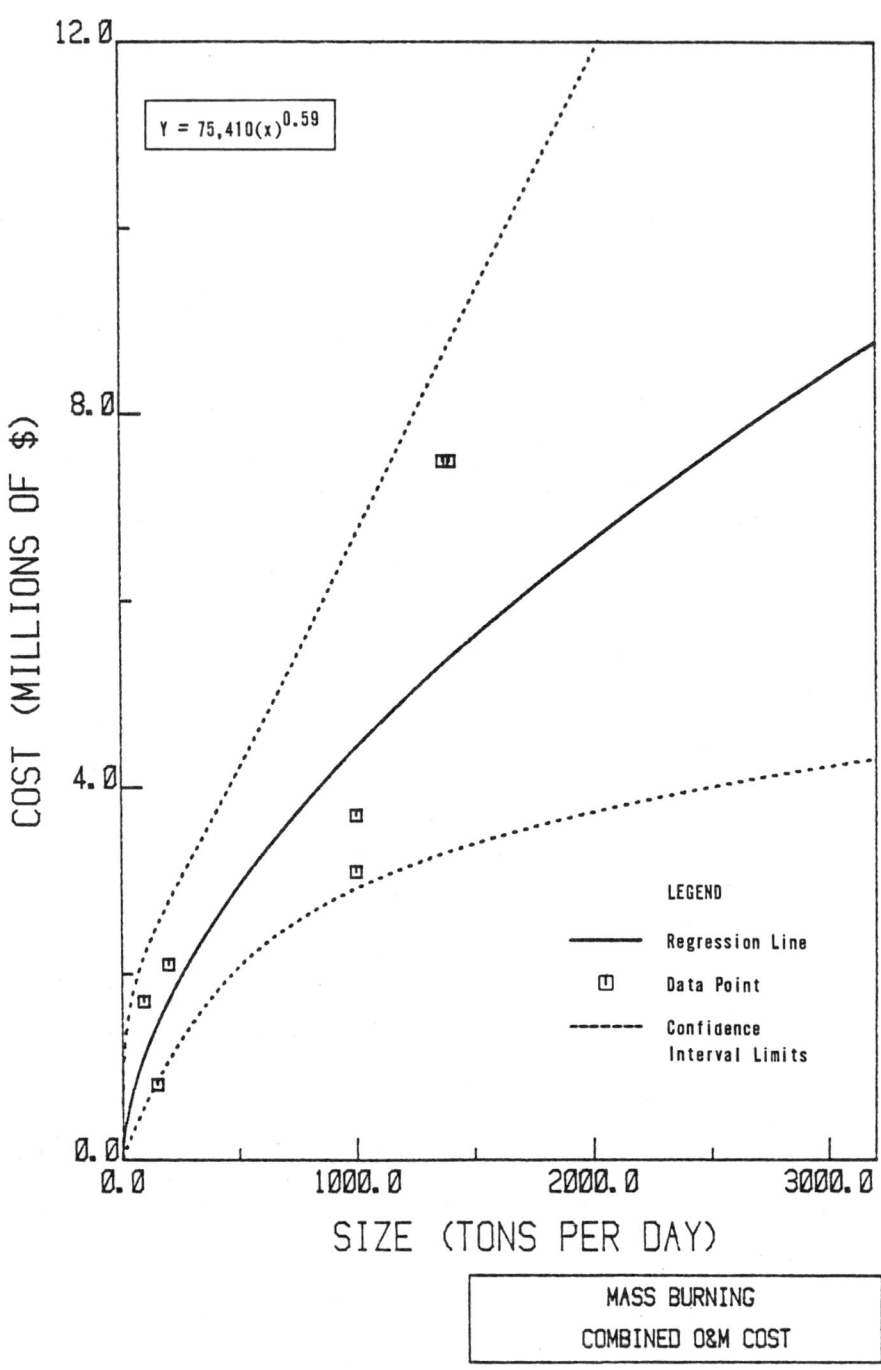

Figure 5-22

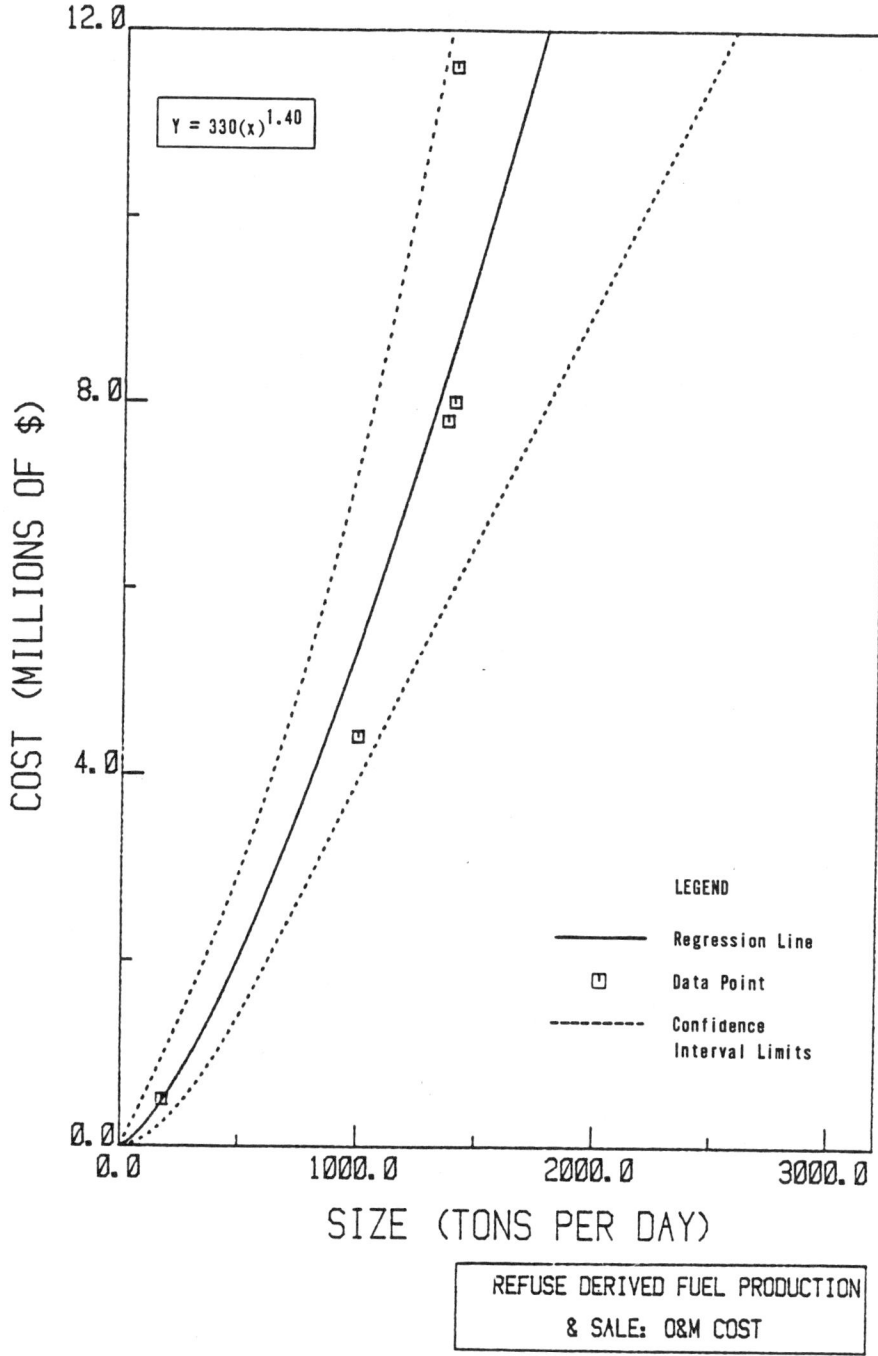

Figure 5-23

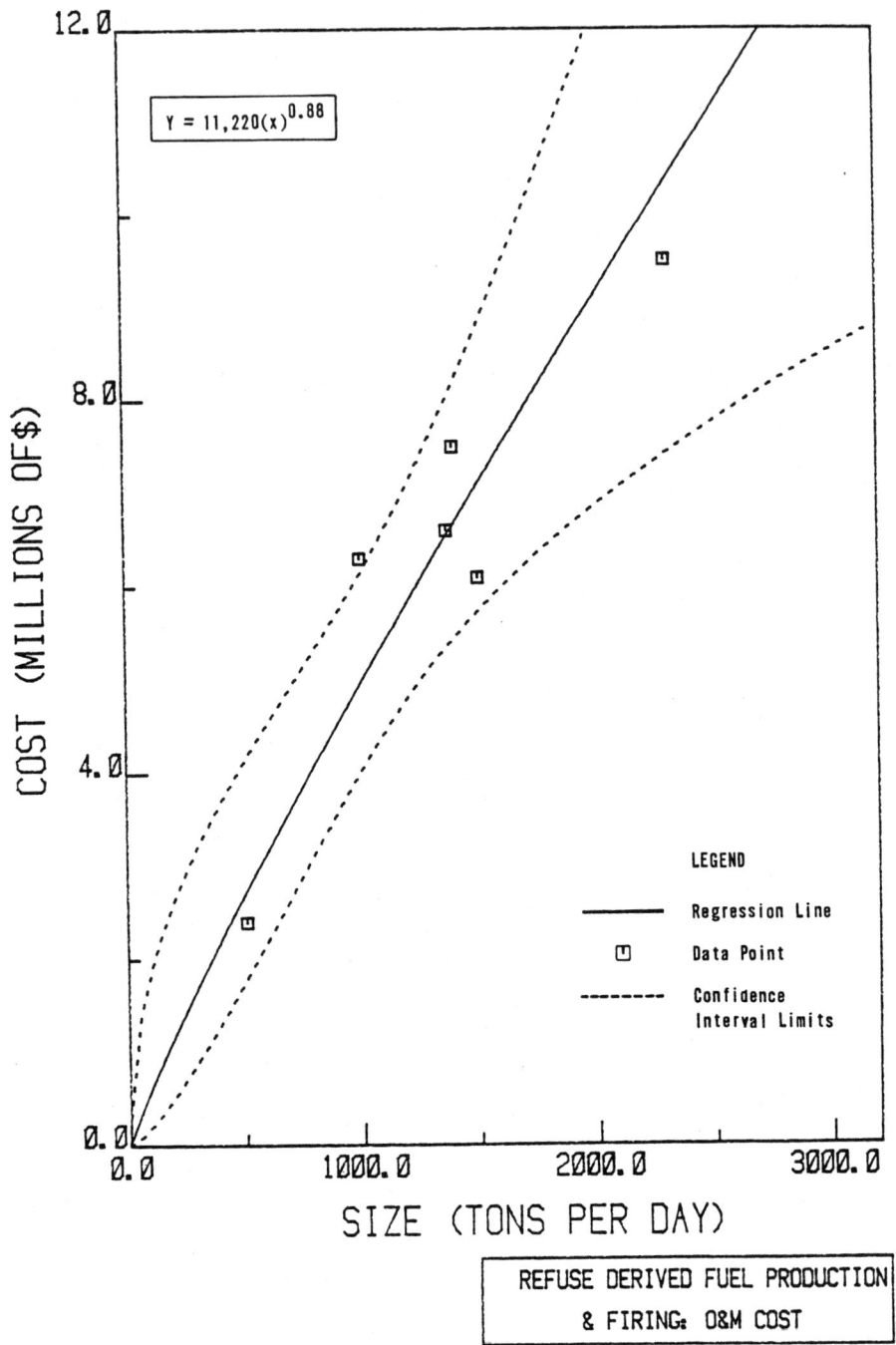

Figure 5-24

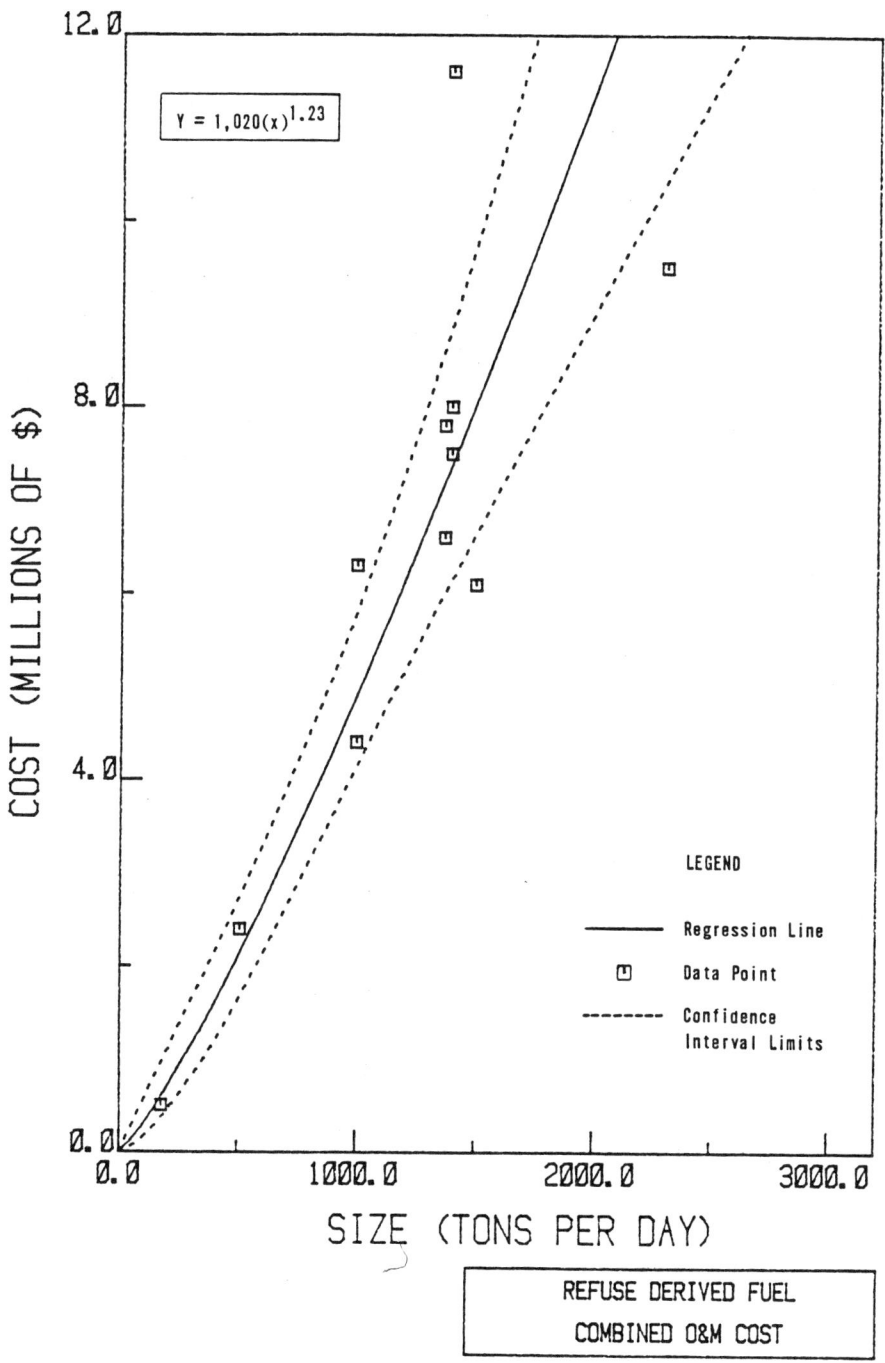

Figure 5-25

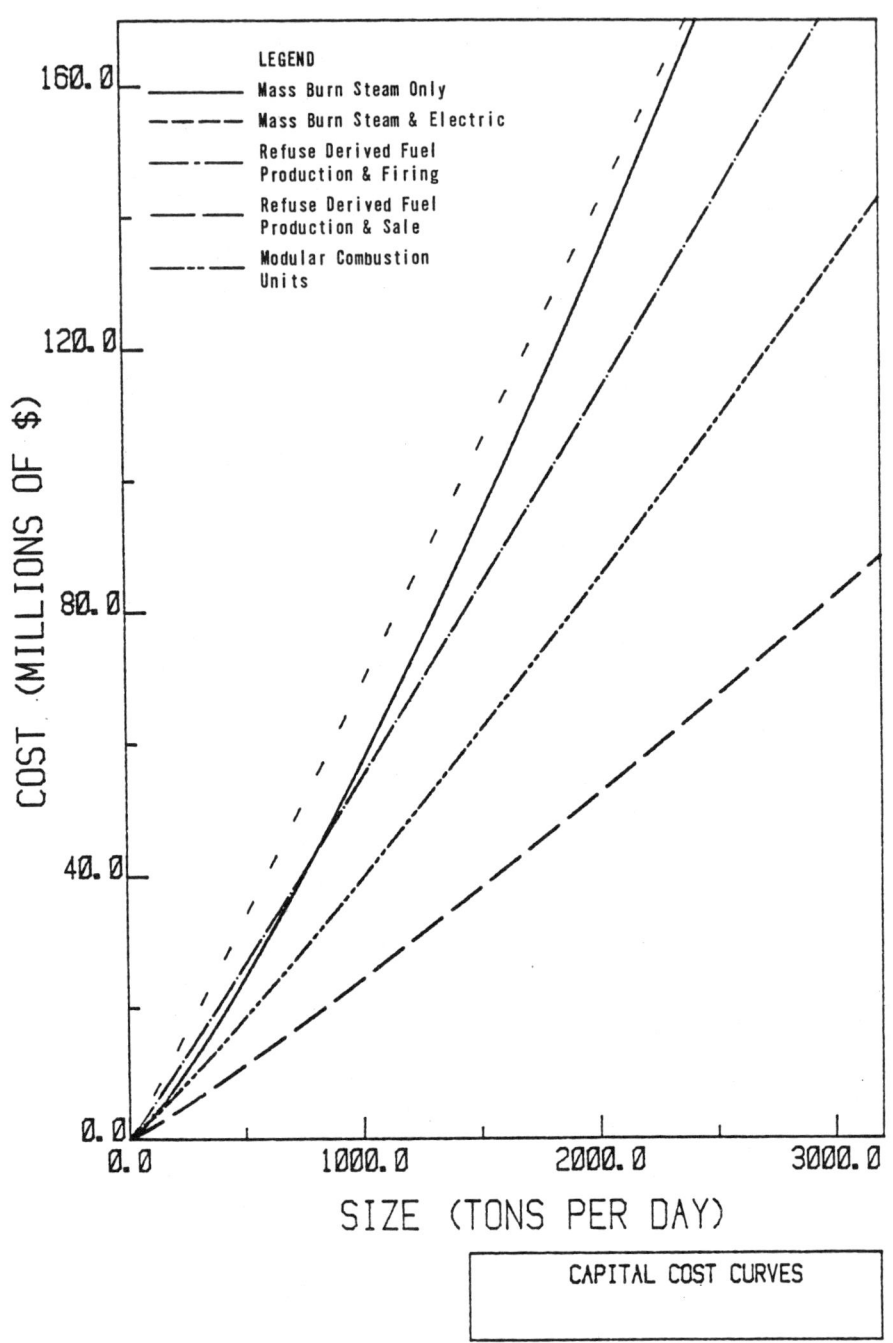

Figure 5-26

Resource Recovery Systems 177

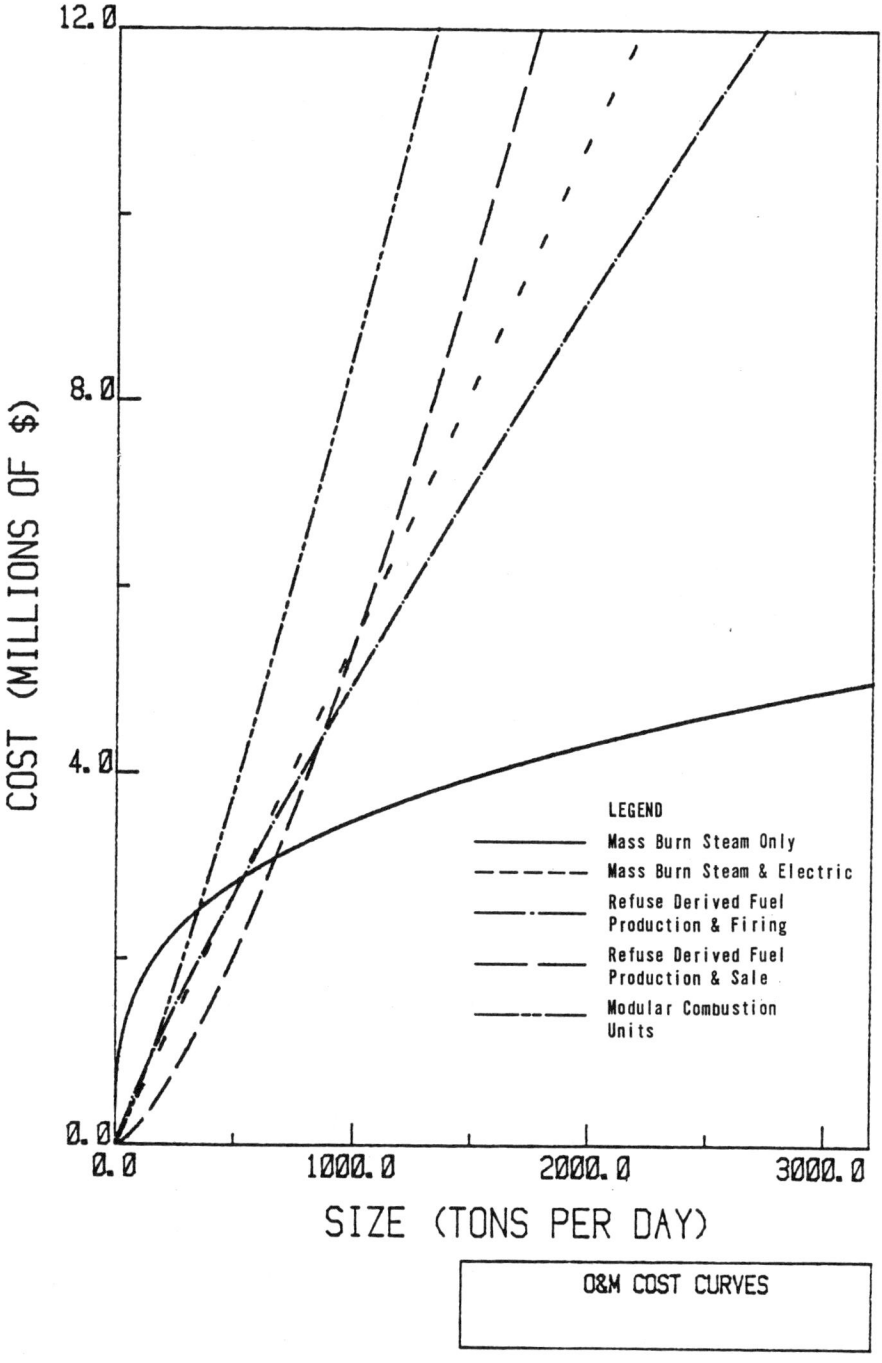

Figure 5-27

REFERENCES

1. Port Authority of NY and NJ, <u>Resource Recovery Installations Data Book</u>, October 1978.

2. Alter, Harvey and J.J. Dunn, Jr., <u>Solid Waste Conversion to Energy</u>, Marcel Dekker, Inc., 1980.

3. Hileman, G.P. and F.B. Pyle, <u>Recyle Energy for Central Heating and Process Steam</u>, presented at International District Heating Association, June 27-29, 1977.

4. Bolton, R.E., <u>An Analysis of Resource Recovery Facility Costs</u>, presented at the ASCE Conference in Environmental Engineering, July, 1982.

6
Codisposal

INTRODUCTION

The combined disposal of solid waste and wastewater treatment sludge is referred to as codisposal. There are many forms of codisposal:

- o Composting of sludge using processed solid waste as a bulking agent.
- o Anaerobic digestion of sludge and the organic fraction of solid waste.
- o Co-pyrolysis of sludge and solid waste.
- o Co-incineration of sludge and solid waste.

This chapter will focus on the last approach, as it is the most widely practiced, and the most data is available on its application.

Wastewater treatment sludge is the residue of the sewage treatment process. The quantities of sludge produced nationwide have increased rapidly in recent years as Federal water pollution control requirements have caused the number and size of sewage treatment plants to increase rapidly.

As of 1980, over 8 million dry tons of sludge were produced a year and this amount can double by 1987.

Incineration is a major disposal technique, as noted on the following page.

Disposal Method		Percent
Thermal Process[1]	1,978,000	23
Distribution-Marketing System[2]	1,806,000	21
Land Application	2,494,000	29
—food chain land	(1,462,000)	(17)
—non-food chain land	(1,032,000)	(12)
Landfill	1,118,000	13
Other[3]	860,000	10
Ocean Dumping	344,000	4
TOTAL	8,600,000	

[1] Primarily incineration, includes pyrolysis.

[2] Sludge that is sold or given away. Includes processing (such as composting or heat drying) to prepare product for market.

[3] Lagoons and/or stockpiles.

Source: Reference 1

EPA has estimated that existing and planned sludge incinerators will use the equivalent of almost two million barrels of fuel oil per year to burn sludge cake. Sharp increases in fuel costs in recent years have lead to a search for cheaper alternative energy sources, including the use of solid waste as a substitute for fuel oil or gas for sludge incineration.

CODISPOSAL TECHNIQUES
General

There are two basic approaches to codisposal by incineration:

- o incineration of sewage sludge in a solid waste incinerator; or
- o the use of refuse derived fuel (RDF) as auxiliary fuel in a sludge incinerator

Incineration with Solid Waste

The earliest technique for co-incineration involved the direct injection of dewatered sludge filter cake[4] into the combustion chamber of a conventional solid waste incinerator. This technique has had a poor record and has generally been abandoned. Problems generally include poor mixing of the filter cake and refuse and incomplete combustion of the filter cake due to crust formation.

One new plant utilizing a variation of this technique is presently under construction at Glen Cove, NY. (see Figure 6-1). At Glen Cove the dewatered sludge is to be extruded as a sheet on deposited above the solid waste entering a conventional mass burning stoker furnace.

Waste heat from the combustion of solid waste can be used sludge prior to incineration.

There are several methods of sludge drying utilizing waste heat:

o direct utilization of hot flue gases;

o direct-contact drying;

o flash drying

o steam drying; or

o use of an intermediate fluid for drying.

Other techniques for the incineration of sludge in a solid waste incinerator include spraying wet sludge directly into the combustion chamber, and use of a flash evaporator to feed wet sludge into a combustor.

Use of RDF in a Sludge Incinerator

There are two common types of sludge incinerators in common use in the United States today: the multiple hearth furnace and the fluid bed furnace.

The Multiple Hearth Incinerator - The multiple hearth incinerator (also known as the Herreshoff furnace) is the most prevalent incinerator for the disposal of sewage sludge in this country. It was developed specifically for sludge burning and has been adapted to carbon regeneration and recalcining (both applications occurring in water treatment) and has numerous industrial applications.

Sludge cake is introduced at the top of the furnace (see Figure 6-2). The furnace interior is composed of a series of circular refractory hearths, one above the other. The hearths are self-supporting off the refractory lines cylindrical wall

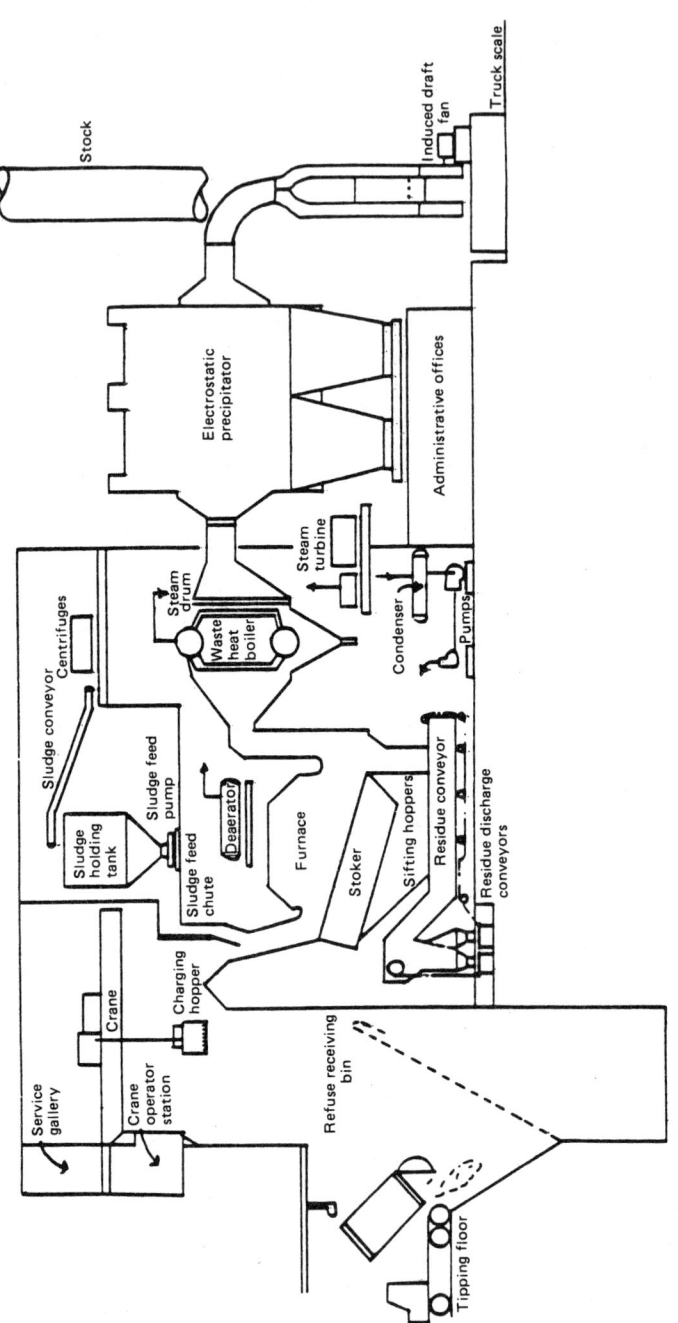

Figure 6-1
The Glen Cove System
(Reference 2)

of the furnace. They are numbered No. 1 as the top hearth, No. 2 as the next to top, etc. There are from five to nine hearths in a typical furnace.

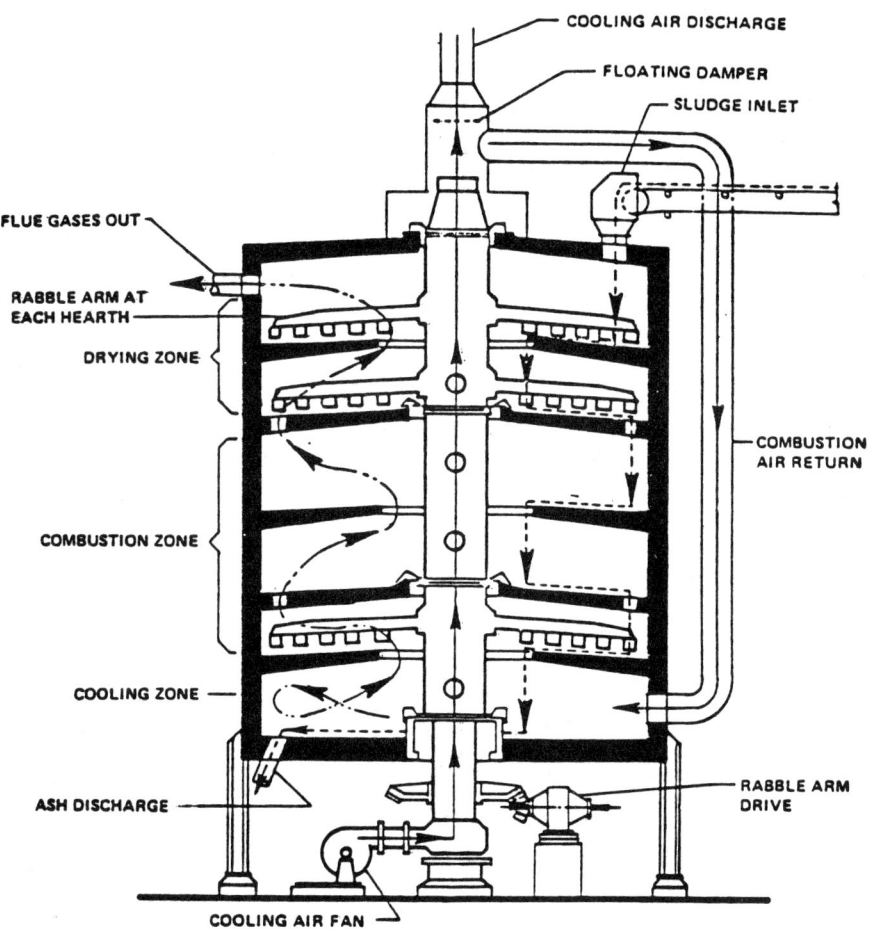

Figure 6-2
Multiple Hearth Incinerator
(Reference 5)

Every other hearth has a large annular opening between the hearth and the center shaft. These are called "in hearths." The teeth on the rabble arms or these hearths will "rabble' sludge to the center of the hearth where the sludge will fall off the edge of the refractory landing on the hearth below, an "out hearth".

There are a series of openings on the outside or periphery of the out hearth. The inside of an out hearth is fairly close to the center shaft. A "lute ring cover" is a collar located above the hearth, attached to and moving with the center shaft, preventing sludge from dropping adjacent to the center shaft. Teeth on the rabble arms on the out hearths move sludge to the drop holes on the outside of the hearth where the sludge drops to the hearth below, an in hearth. This process repeats until sludge, or ash, reaches the bottom hearth, the floor of the furnace, where it discharges from the furnace.

Teeth on each hearth agitate the sludge, exposing new surfaces of the sludge to the gas flow within the furnace. As sludge falls from one hearth to another, it again has new surfaces exposed to the hot gas. The upper hearths of the furnace comprise a drying zone where the filter cake gives up moisture while cooling the hot flue gases. Flue gas exits the top hearth of the furnace at from 800° to 1000°F. The center hearths are the burning zone where temperatures can reach 1700° to 1800°F. Burn out of sludge to ash is accomplished in the lower hearths of the furnace.

The center shaft and rabble arms are hollow. Air is passed through the center shaft, which is constructed to distribute this air to each of the rabble arms, and discharge it through the top of the shaft. This air is utilized for shaft cooling air and after passing through the center shaft, it reaches temperatures of 200° to 450°F. Often this heated air is recycled into the furnace as preheated combustion air.

Excess air of 100 to 125% must be provided to insure adequate burn out of sludge. Some 20% of the ash content of the sludge is airborne and gas cleaning equipment must be provided for its capture. Occasional odor problems will exist which may require installation of after-burning equipment.

To reduce the auxiliary fuel costs of operating a MHF, development work has been done operating a MHF in excess and starved air modes, using RDF. In 1975, the Central Contra Costa Sanitary District in California conducted a full scale 45 day demonstration project on a 16 ft multiple hearth furnace, using a mixture of sludge and RDF. The test was generally successful and demonstrated that autogenous combustion could be maintained with an RDF/sludge weight ratio of 1:2, at a sludge solids content of at least 16 percent.

When operated in a starved air mode, the MHF produces a fuel gas with a heating value in the order of 130 Btu per cu ft. This gas is fired in an afterburner and it has the potential to produce recoverable energy.

Fluidized Bed Incineration - The fluid bed furnace was developed for catalyst recovery in oil refining by Standard Oil early in this century. The first fluid bed used for incineration of sewage sludge was installed in 1962. Its use is gaining in popularity in the United States.

As illustrated in Figure 6-3, the fluid bed furnace is a cylindrical refractory lined shell with structure on its bottom surface to support a sand bed.

Air is introduced at the fluidizing air inlet at pressures in the order of 3.5 to 5 psig. The air passes through openings (tuyeres) in the grid supporting the sand and creates fluidization of the sand bed.

Air can be introduced cold, or as usually the case, preheated by the exiting flue gas. The sand bed is maintained at approximately 1500°F. It expands 30 to 60% in volume when fluidized as compared to the unfluidized condition.

Sludge cake is normally introduced within the fluid bed. The fluidizing air flow must be carefully controlled to prevent the sludge from floating on top of the bed. Fluidization provides maximum contact of air with sludge surface for optimum burning. The drying process is practically instantaneous. Moisture flashes into steam upon entering the hot bed.

The furnace itself is an extremely simple piece of equipment with no moving parts. The large amount of sand within the furnace is an effective heat sink. The furnace can be shut down with minimal heat loss. It is a relatively tight system and the sand will retain heat to allow startup after a weekend shutdown with need for only one or two hours of heating. The sand bed should be at least 1200°F before sludge is introduced.

Because of the intimate mixing of air and sludge in the fluid sand bed, excess air requirements are low, from 20 to 40%. The large volume within the furnace above the sand bed is maintained at 1200° to 1500°F. Residence time of the flue

186   Energy and Resource Recovery from Waste

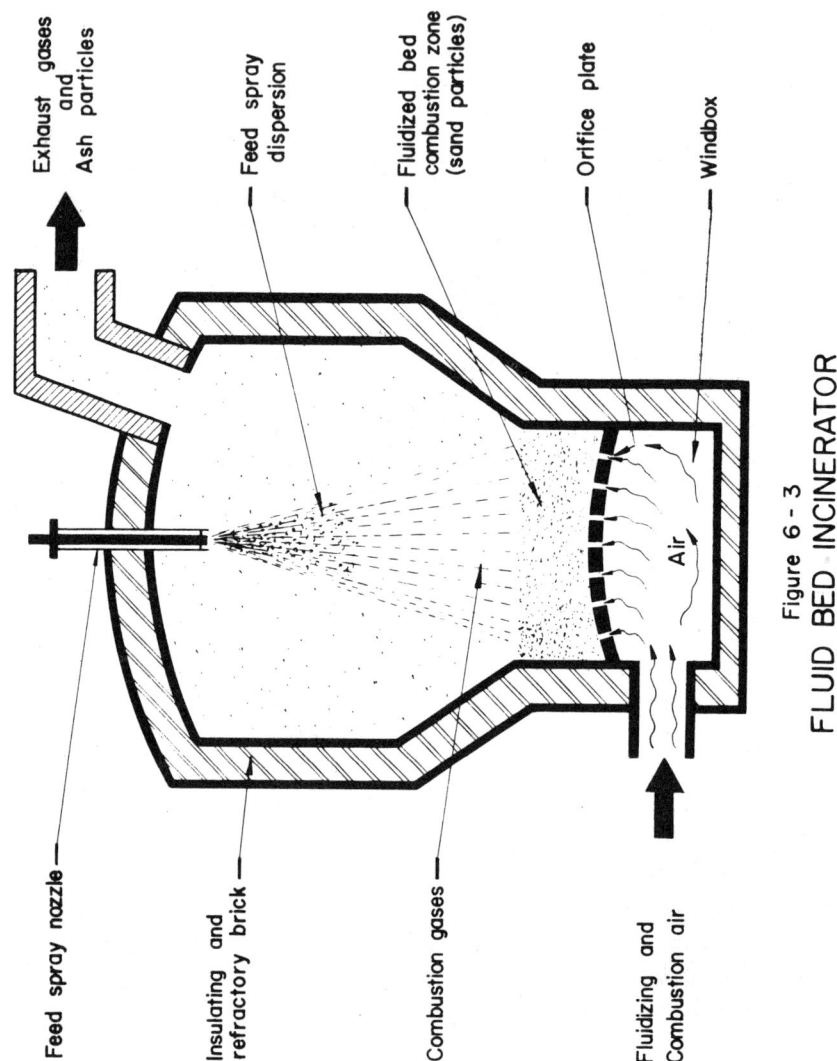

Figure 6-3
FLUID BED INCINERATOR

gases at these temperatures is sufficient to obtain complete burn out and elimination of odors.

Sludge is force-fed into the furnace with either positive displacement pumps or screw/plunger type feeders. Sludge feeding has had its problems because of the tendency of the sludge within the feeder to dry and harden during periods when the furnace is maintained hot without sludge feed. (Hot stand-by conditions).

Sand has to be made up at the rate of approximately 5% of the bed volume every 300 hours of operation when burning sludge cake.

Fuel is used for startup, reheat, and depending on the properties of the sludge, for incineration. It can be injected within the bed or sprayed on top of the bed.

The fluid bed furnace has one major item of air moving equipment, the forced draft fan (or fluidizing air blower). The fan is sized to blow the flue gas through the gas scrubbing systems which necessitates that the reactor be pressurized and that it be tight to prevent leakage of flue gas.

Processed solid waste and sewage sludge have been incinerated in a FBF as part of the Black Clawson demonstration project in Franklin, OH. This process utilizes a wet-pulp RDF described more fully in Chapter 5. This project is no longer in operation.

A second system of this type, utilizing a dry RDF system, is in operation at the Western Lake Superior Sanitary District sewage treatment plant at Duluth, MN.

CURRENT STATUS

Table 6-1, adapted from Reference 3, summarizes the current status of codisposal projects in the United States and Europe. It can be seen that although codisposal has proven successful in Europe, there are few successful facilities in the US.

Some of the reasons for lack of success of codisposal facilities in the US include:

o Many of the incinerators in which it was tried were not originally designed for codisposal;

o Many of the incinerators in which it was tried had out of date pollution control equipment and encountered emission problems;

TABLE 6-1
CODISPOSAL OF MUNICIPAL SOLID WASTE
AND SEWAGE SLUDGE

	Technique	Location	Remarks
1.	Incineration of dewatered sludge filter cake and raw solid waste in a conventional solid waste incinerator.	Kewaskum, Wisconsin	Abandoned
		Whitemarsh, Pennsylvania	Abandoned
		Cheneviers, Switzerland	
		Frederick, Maryland	Abandoned
		Glen Cove, New York	Under Construction
2.	Incineration of dewatered sludge filter cake in a multiple-hearth unit employing raw solid waste as an auxilliary fuel.	Reigate, England	Lurgi System
		Ebingen, Germany	Lurgi System
		Bulach, Switzerland	Lurgi System
		Dubendorf, Switzerland	Nichols System
		Bowhouse, Alloa, Clackmannshire, Scotland	Nichols System
3.	Combustion of dewatered sludge filter cake with refuse derived fuel in a fluidized-bed incinerator	Franklin, Ohio	Special; not in service
		Menlo Park, California	Demonstration Plant
		Lausanne, Switzerland	Tests Only
		Great Lakes Paper Co., Thunder Bay, Ontario, Canada	Tests Only
		Duluth, Minnesota	In Startup
4.	Utilization of waste heat from the combustion of raw solid waste to evaporate moisture from wet sludge (10 percent solids) prior to incineration of sludge with refuse in the same combustion chamber.	Ansonia, Connecticut	Spray Dryer; not in service
		Holyoke, Massachusetts	Rotary Dryer; not in service
		Dieppe, France	von Roll System
		Stamford, Connecticut	Rotary Dryer; not in service
		Gluckstadt, Holstein West Germany	Drag-Conveyor Dryer
		Krefeld, Germany	Flash-Drying System (probably)
		Harrisburg, Pennsylvania	
		Nurnberg, West Germany	von Roll System; proposed
		Lulea, Sweden	Rotary Dryer

(continued)

TABLE 6-1
(continued)

Technique	Location	Remarks
5. Utilization of spray techniques to inject wet sludge directly into the combustion chamber of a refuse incinerator.	Alrincham, United Kingdom	Abandoned
	Dickerson Station, Havant, United Kingdom	Proposed
6. Utilization of flash-evaporation techniques to feed wet sludge into a solid waste combustor.	Neenah-Menasha, Wisconsin	Shut down
	New Albany, Indiana	Shut down
	Waterbury, Connecticut	On standby
	Watervliet, New York	Shut down
	Bloomsburg, Pennsylvania	Abandoned
	Essen Karnap, West Germany	Proposed
	Newburgh, New York	Abandoned
	Eastman Kodak, Rochester, New York	Special
	Louisville, Kentucky	Not co-incinerating
	Trenton, Michingan	Drying only
7. Pyrolysis	Memphis, Tenn.	Multiple Hearth Pryolysis; Under Design
	South Charleston, W. Virginia	Shut down
	Baltimore, Maryland	Monsanto System; not co-incinerating
	Orchard Park, New York	Torrax-co-incineration; tested
	San Diego, California	Garrett System; abandoned
	Kalnudborg, Denmark	Tital Thermogen System; inactive refuse only
	Minneapolis/St. Paul, Minn.	Proposed
	Contra Costa, California	Multiple Hearth pyrolysis; Facility plan complete

(continued)

TABLE 6-1
(continued)

Technique	Location	Remarks
8. Miscellaneous	Glouchester City, New Jersey	Abandoned
	International Paper Company Mobile, Alabama	Sludge & Hogged Fuel
	Georgetown, South Carolina	
	Vicksburg, Mississippi	
	Manistique Pulp and Paper Co. Manistique, Michigan	
	Farberjabriken Baya A.G., Laverkusen, Germany	Rotary Kiln
	Buick Motor Division, GMC, Flint, Michigan	Oily Sludge only
	Plaine de Rhone, France	MacLaren System; proposed
	Pleasantville, New Jersey	Salt bath
	WIBAU Matthias Plant, South Germany	
	Riegal Paper Company Milford, New Jersey	Rotary Kiln
	Cologne, West Germany	Multiple Hearth
	Dordrecht, Netherlands Keller-Peukert System	Conceptual

Source: Reference 3

o Inter-jurisdictional problems often developed between wastewater treatment and solid waste agencies; and

o When energy costs were lower and land disposal inexpensive and available, there was little incentive to make codisposal work.

Table 6-2, adapted from Reference (5), lists technical consideration for furnace conversion to codisposal, both Multi-hearth (MHF) and Fluidized (FBF).

TABLE 6-2

TECHNICAL CONSIDERATIONS FOR RDF USE IN EXISTING SLUDGE INCINERATORS

Consideration	MHF-Pyrolysis	FBF
RDF process	Shred, air classify, screening 65% RDF output d-RDF may be preferable	Trommel, coarse air classify
RDF specification	<2-3" particle size Low inerts Low Btu/lb and moisture content variation preferred	Coarse 95% 6" Low inorganic content preferable Btu and moisture varia- not critical
Feeding	Screw conveyor feed at 1 or 2 points	Ram feeding or pneu- matic feeding Feed into sand bed
Site	Individual analysis re- quired Waste heat boiler location should be as close as possible	Individual analysis required Waste heat boiler location should be close as possible
Fuel usage	Significant supplementary fuel required for warm- up periods	Due to heat sink, nature of design, less supplementary fuel required at full loads
Incinerator modifications	Add feed system Replace induced draft fan Seal air leakages Water jacketed ash screw Afterburner required	Add waste heat boilers especially to older installations Add bed-removal clas- sification system if RDF quality is low

(continued)

TABLE 6-2
(continued)

Consideration	MHF-Pyrolysis	FBF
Incinerator additions	Add and raise rabble arms Additional controls and instrumentation Add waste heat boiler Anticorrosion lining for off-gas system Add split-draft ducting	Modify/add scrubber capacity Make "hot shell" design in order to inhibit corrosive gas condensation of HCl and H_2SO_4 Add feed system
Cost to convert	$1-2M per conversion, depending upon unit size for boiler, air pollution modifications, etc.	$0.15-0.5 M per conversion depending upon size for boiler, air pollution modifications, etc.
Ash handling	Ash output increased, has larger heat sink, requires water jacketed ash removal system	With high inorganic RDF, bed material will require removal and classification Ash at scrubber and waste heat boiler will increase
Air pollution	Increase in scrubber capacity required	Increase in particulate emissions and shift in particulate size
Operational	Around-the-clock operation will result in lower maintenance costs and longer equipment life	Intermittent operation less harmful due to heat sink nature of sand bed material
Waste heat recovery potential	60-75% of input BTU's	50% of input BTU's

ECONOMICS

Capital costs for codisposal facilities are high compared to land disposal alternatives such as sanitary landfill or landspreading.

Nevertheless, many studies have shown codisposal to be economically feasible when compared to separate sludge and solid waste facilities. The keys are:

o a market for recovered energy;

o the ability to charge a reasonable tipping fee; and

o EPA funding.

In the past, EPA funding policy on codisposal projects has been confused and restrictive. In general, municipal wastewater treatment projects are 55 to 75 percent funded by the Federal Government under the Construction Grants Program. Construction of solid waste facilities is generally not funded. For multipurpose projects EPA policy is to fund pro-rated costs and fund only those specifically pertaining to sludge. This can act as a disincentive to codisposal projects, since a sludge-only alternative may result in greater Federal funding, even though the long range costs of the codisposal option may be lower.

REFERENCES

1. JRB Associates, *Solid Waste Data; A Compilation of Statistics on Solid Waste Management Within the United States*, USEPA, PB 82-107301, 1981.

2. U.S. General Accounting Office, *Codisposal of Garbage and Sewage Sludge -- A Promising Solution to Two Problems*, 1979.

3. Roy F. Weston, Inc., *A Review of Techniques for Incineration of Sewage Sludge with Solid Wastes*, EPA-600/2-76-288, 1976.

4. Gordian Associates, *Assessment of the use of Refuse-Derived fuels in Municipal Wastewater Sludge Incinerators*, 1977.

5. Brunner, C.R., *Design of Sewage Sludge Incineration Systems*, NOYES Data Corp., 1980.

7
Source Separation

INTRODUCTION

Source separation is the setting aside of recyclable waste materials (such as paper, glass, and metal containers) at their point of generation (the home, office, or other place of business) by the generator.

Source separation of paper, which is the largest component of solid waste, has been widely practiced in the United States. Separate collection of cans and bottles is less common, and less data is available. This Chapter will therefore focus on separation and recycling of paper, although the methods discussed here are equally applicable to other recyclable materials.

It is important to note that source separation is not by itself a complete solid waste disposal method. As Example 7-1 shows, the maximum source separation that might reasonably be expected is six to seven percent of the total solid waste flow.

In addition, the solid wastes remaining after source separation tend to be higher in moisture content, making disposal by landfill or incineration more difficult. Nevertheless, source separation can make a valuable contribution as part of an overall solid waste management and disposal program.

EXAMPLE 7-1

Calculate the maximum contribution of source separation to solid waste disposal: Consider the solid waste composition given in Table 2-4.

Paper	40% by weight of refuse
Cardboard	4
Ferrous Metals	8
Nonferrous Metals	1
Glass	8
	61% of refuse

Of this 61% not all is potentially recyclable.

Newspaper, 20-25% recovery or	9% by weight of refuse
Cardboard, up to 100% recovery or	4
Ferrous Metals, up to 75% recovery or	6
Nonferrous Metals, approx. 50% recovery or	0.5
Glass, up to 75% recovery or	6
	25.5% of refuse

Municipal solid waste is between 50 and 60 percent of total solid waste: the balance is industrial, construction and demolition debris, cars, trees, sludges, and so forth. Thus source separation can, <u>at most</u>, handle 25 percent of 50-60 percent or:

25% x 50-60% = 13 - 15% of the total

However, no community has achieved 100 percent participation in a source separation program. If 50 percent participation is assumed; then:

13 - 15% x 50% = 6-7%

COLLECTION METHODS

General

The two basic methods for separate collection involve the use of separate vehicles, or the use of a rack attached to a standard refuse collection truck.

Separate Truck

Standard packers, vans and open-bodied trucks have all been used for separate collection. Packers are preferred, as other vehicle types require an extra crew member to stack paper inside the truck.

The separate collection truck can generally cover three to five normal routes each day because:

- there are fewer items to handle at each stop;
- there are no containers to be returned to the curb;
- not all households participate; and
- not every participating household places paper at the curb every collection day

Collections are usually monthly, by-weekly, or weekly. EPA data indicates higher participation rates for higher frequencies of collection, although this is offset to some extent by higher collection costs.

Costs for implementing these systems include capital, operating and maintenance costs for the truck, as well as the labor costs for additional personnel. In many cities, a truck from a standby fleet can be used, eliminating extra capital cost. It may also be possible to avoid employment of additional labor. May solid waste collection personnel do not work a full 40 hour week, due to "incentive" programs (crews go home when they finish their route), four day a week pickup schedules, and other factors. They are often available for overtime work and paying them overtime wages for separate pick-up may be cost-effective compared to hiring additional personnel.

<u>Rack System</u>

In this system racks, varying in capacity from ½ to 1½ cubic yards, are installed underneath the body of a standard packer truck (see Figure 7-1). Bundled newspaper and separately bagged cans and bottles are then placed at the curb by the homeowner, along with ordinary refuse, thus avoiding the need to have a separate collection day.

The rack method has the advantage that each route must be covered only once, when regular collection is performed. A disadvantage is the tendency for the racks to become filled before the body of the truck has been filled with ordinary refuse. This wastes time because the refuse crew must interrupt collection to unload the racks, either at bulk containers placed along the route, or by traveling to a separate disposal point, or recyclable materials may be dumped into the truck.

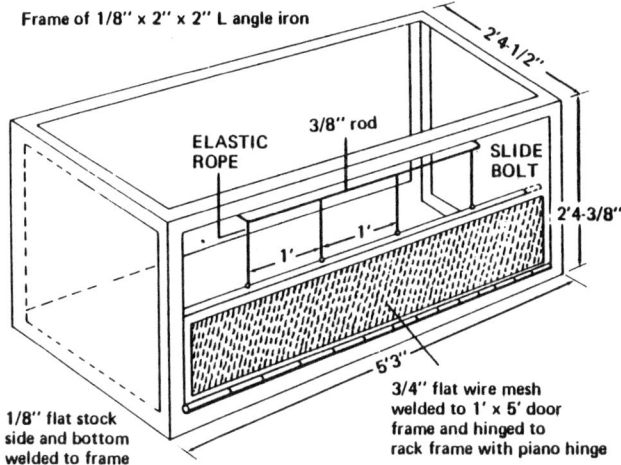

Figure 7-1
Newspaper racks for packer trucks
(Madison Wis., Department of Public Works)

Other methods of collecting newspapers separately with a modified packer truck are under development, including a two-compartment truck, use of an elevated rack, and other devices. These developments could improve the economics of separate collection, if proven feasible.

The rack collection system adds about $250 to the cost of a standard collection vehicle. In addition, between 10 and 14 seconds extra are required at each stop to load newspaper, and 5 to 15 minutes are required to unload the racks when filled. Although additional labor may not be required if collection personnel are not now working a full day, the system must be able to absorb the additional time requirements, or the revenue from paper must be able to offset additional costs.

COSTS

Due to the large number of variables, it is difficult to determine average costs for separate collection. Among the factors which affect economic feasibility are:

- o type of collection practiced (frequency, crew size, etc.);

- o alternative disposal costs;

- o revenue received for recovered materials;

- o participation rate;
- o availability of underutilized men and equipment; and
- o ability to take advantage of reduced waste volumes to reroute regular vehicles

Table 7-1, derived from Reference 1, summarizes collection and disposal costs for 10 communities, both before and after separate collection programs were instituted. While some cities decreased overall costs with source separation, others experienced an increase. The revenues from the sale of newspaper obviously have an important effect on economics: during a period of high newspaper prices the 10 cities experienced an average 5 percent savings with source separation. When newspaper prices were low source separation raised prices for the same cities by 2.6 percent.

TABLE 7-1

COSTS OF SOURCE SEPARATION

Case Study Location	Collection and Disposal Cost Prior to Separate Collection	Collection and Disposal Cost With Separate Collection*			
		Low Paper Market (average $8 per ton)		High Paper Market (average $25 per ton)	
	($/Ton)	($/Ton)	(% Change)	($/Ton)	(% Change)
Dallas, Tex.	12.10	11.60	-4.1	9.30	-23.1
Fort Worth, Tex.	13.50	14.10	+4.4	11.80	-12.6
Great Neck, N.Y.	36.00	38.70	+7.5	36.50	+1.4
Green Bay, Wisc.	38.70	37.70	-2.6	37.10	-4.1
Greenbelt, Md.	27.20	27.40	+0.7	26.30	-3.3
Marblehead, Mass.	23.10	25.30	+9.5	24.10	+4.3
Newton, Mass.	32.40	32.20	-0.6	31.60	-2.5
University Park, Tex.	14.70	14.90	+1.4	13.10	-10.9
Villa Park, Ill.	13.50	13.40	-0.7	12.40	-8.1
West Hartford, Conn.	26.30	26.50	+0.8	25.20	-4.2

* Credit given for diverted disposal costs and revenue generated from the sale of separately collected wastepaper.

Example 7-2 shows how the costs of source separation can be calculated for a specific application.

EXAMPLE 7-2

Consider a town of 50,000, generating 3 lbs/cap/day solid waste, utilizing landfill disposal at $15 per ton. If the town institutes a source separation program utilizing standard trucks equipped with 4 cu yd racks, and sells the newspaper collected for $25 per ton, what are the net revenues (or costs)?

Assume each family is 4 persons. Then:

4 x 3 lb/cap/day x 7 = 84 lbs/wk/house

Assume 2 pickups/wk. Then:

84/2 = 42 lbs/pickup

Assume each pickup takes 0.55 minutes from stop to stop. Then the average loading rate is:

0.55/42 x 2000 = 26.2 mins/ton

Assume 20 cu yd trucks, average load 5 tons. Then to fill one truck takes:

5 x 26.2 = 130.9 mins

Assume that the landfill is 30 minutes roundtrip from the collection routes. Consider 1 load/truck/day:

Time available	8 x 60 =	480 minutes
less Lunch		60
less Other non-collection time		60
less Landfill trip		30
less Fill truck		131
Total Unoccupied Time		199 minutes

Since 199 minutes are available at the end of the day and only 131 minutes are required to load a truck, two loads/truck/day will be considered:

Time available	480 minutes
less Lunch	60
less Other non-collection time	60
less 2 landfill trips	60
less Fill truck twice	262
Total Unoccupied Time	38 minutes

Conclusion: Each truck can handle 2 loads a day.

Calculate the total number of trucks required:

Total generation = $\frac{50,000 \times 3 \times 7}{5}$ = 210,000 lbs/collection day

= 105 tons/collection day

Since each truck carries 5 tons x 2 trips/day = 10 tons per collection day:

    105/10 = 10.5   say 11 trucks required

Now consider impact of separate newspaper collection. Assume separate collection adds 15 seconds to each pickup:

    0.55 + (15/60) = 0.80 minutes/pickup

    0.80/42 x 2000 = 38.1 min/ton

Each newspaper rack holds 4 cu yds. Assume 150 lbs/cu yd density:

    (4 x 150)/2000 = 0.3 tons
    Total truck capacity = 5 + 0.3 = 5.3 tons
    Time to fill = 5.3 tons x 38.1 min/ton = 201.9 min

Now, consider one load per truck per day:

Time available	480 minutes
less Lunch	60
less Non-collection	60
less Landfill trip	30
less Fill truck	202
Total Time Unoccupied	128

Since 128 minutes is less than the 202 minutes required for filling the truck, each truck can handle only one load a day. The total number of trucks required is:

$$\frac{105 \text{ tons/collection day}}{5.3 \text{ tons/truck}} = 19.8 \qquad \text{say 20 trucks}$$

Therefore, separate collection requires:

$$20 - 11 = \underline{9 \text{ extra trucks}}$$

Assume that the total cost of operating a collection truck is $25 per hour. Then the operating cost of the extra trucks is:

    9 trucks x $25 x 260 days x 8 hours/day = $468,000/year

Assume that newspaper is 10% by weight of the total solid waste, and that 50% participation in source separation is achieved. Then newspaper revenues are, at $25 per ton:

    105 x 0.10 = 10.5 tons per collection day generated
    10.5 x 0.5 = 5.3 tons per collection day collected
    5.3 x 260 x $25/ton = $34,450/year

Check to make sure that collection is not limited by rack capacity:

    0.3 tons/rack x 20 truckloads/collection day = 6 tons/collection day

6 tons is greater than the anticipated 5.3 tons collected paper. Therefore, rack capacity does not limit collection.

Calculate saving at landfill:

5.3 tons/C-D x $15/ton x 260 = $20,670/year

Calculate net cost:

Extra trucks	$468,000/year
Less newspaper revenues	34,450
Less landfill savings	20,670
Net cost	$412,880/year

This cost does not include the capital cost expenditure for new trucks. Also, by adjusting pick-up routes less trucks may be required than the 9 extra trucks calculated.

PUBLIC EDUCATION

The success of a source separation program depends heavily on public awareness and cooperation which can only be developed by a vigorous public education program to explain the goals and methods of the program. Television, newspapers, billboards, flyers, and oral presentations have all been used successfully to inform the public. In addition, local environmental and service organizations are often willing to lend their active participation.

USEPA has issued a brochure which describes how a community can plan, implement, and sustain an ongoing publicity campaign. See Reference 2.

REFERENCES

1. USEPA, *Residential Paper Recovery, A Municipal Implementation Guide*, 1975.

2. USEPA, *Residential Paper Recovery*, A Community Action *Program*, SW-553, 1976.

8

Environmental Considerations

GENERAL

The operation of a resource recovery facility produces a variety of environmental impacts, most of which are typical of light to medium industrial facilities. These impacts, including construction, operation, and siting considerations should be given careful attention in the planning process.

Even though several resource recovery technologies have proven viable in the United States over the past ten to fifteen years, most of these facilities have not received widespread attention regarding their methods of operation or their relatively minor impact on surrounding areas. On the other hand, a great deal of local and regional publicity has been associated with the few facilities that have not operated successfully and which have caused detrimental environmental impacts. This emphasis on failure has lead to a misconception in the mind of the average citizen, one that equates resource recovery facilities with open dumps or conventional incinerators. As a result, the attempted establishment of a resource recovery facility in or near a populated area very often encounters intense public opposition. In most cases, acceptance of these facilities can be achieved through a public information program designed to inform the citizenry about the technology to be employed and potential environmental impacts, both beneficial and adverse, associated with it. Generally, the earlier such a program is initiated in the facility planning process, the greater are its chances for success.

Because of the importance of public acceptance in the selection of a site for establishment of a resource recovery facility, several central issues around which public opposition usually centers, should be kept in mind. These factors will also be the focus of a comprehensive environmental evaluation of the facility:

- o the facility must be readily and easily accessible to solid waste transport vehicles;

- o solid waste must be unloaded and stored until processed;

- o solid waste must be moved to some degree during processing;

- o virtually all resource recovery combustion technologies produce air and/or water pollutants during processing;

- o unmarketable residues produced must be stored and/or transported, and finally disposed of in an environmentally sound and economically acceptable manner; and

- o the facility will, at a minimum, have an aesthetic impact on surrounding land use during construction and operation.

While the above listed factors emphasize negative aspects of resource recovery, such negative aspects tend to be the primary public issues. Resource recovery results in several beneficial environmental impacts that should also be emphasized, though not misrepresented, through a public information program. Such impacts should also be fairly addressed during environmental evaluations. The following are such beneficial impacts associated with resource recovery facilities:

- o resource recovery technologies supplement or can largely replace existing disposal methodologies that are more resource utilization intensive;

- o through the recovery of materials and/or energy, both economic and resource costs associated with solid waste disposal are offset to some degree;

- o resource recovery represents long-term disposal capability not provided by many alternative methods; and

- o the recovery of materials and/or energy in a readily

usable form often provides the resource needed, in the area of the facility, for further economic development.

The following portions of this Chapter address specific environmental issues associated with resource recovery. Most, if not all, of these issues are environmental concerns associated with resource recovery regardless of the technology employed.

FACILITY SITING

Of all the decisions associated with facility planning, the selection of a site has perhaps the greatest long-term impact and is certainly the issue having greatest potential for public controversy. Basically, the magnitude and extent of many of the environmental impacts from such a facility are directly affected by the nature of the selected site and its surrounding environs. The sensitivity of the proposed site to potential environmental impacts will, to a large extent, determine the significance of the facility's environmental effects and its public image as a good or bad neighbor. Several basic factors should be considered during the analysis of alternative sites to identify their degree of sensitivity. These include:

- o the accessibility of the site for entry and exit of solid waste transportation vehicles and the type of land use along roadways leading to the site entrance;

- o the type of land use found in areas adjacent to the site, especially the degree of residential development;

- o the proximity of the site to sensitive potential receptors of impacts such as schools, churches, nursing homes, and hospitals;

- o if air pollutants are generated by the process, the quality of ambient air in the immediate region, especially with regard to USEPA established criteria air pollutants (particulate matter, sulfur dioxide, nitrogen oxides, carbon monoxide, photochemical oxidants and lead);

- o if water pollutants are generated by the process (including heated cooling water), the quality of surface waters to which the pollutants will be discharged and the uses of the surface water body, e.g., for recreational purposes and by its biotic inhabitants (i.e., a spawning or nursery area for fish);

o the location of the site in terms of centralization of the wasteshed and proximity to major collection areas and transfer stations,

o the proximity of the site to the facility designated for disposal of unmarketable residue generated by the process;

o the proximity of the site to recycling facilities capable of using materials recovered and/or facilities capable of using recovered energy (steam, electricity, or prepared solid fuel); and

o the existing use of the proposed site, especially with regard to its current or potential use by other industrial or commercial facilities and its natural or cultural resource characteritics (e.g. prime agricultural land, wetland, critical wildlife habitat, parkland, historic buildings, or archeological remains).

These factors should be used to screen alternative sites being considered for location of a resource recovery facility. Public involvement should be solicited during this screening process in order to keep the citizenry informed during facility planning and to help determine those issues of greatest local concern.

Because resource recovery as a disposal method is most often associated with urbanized wastesheds, the availability of suitable sites is frequently limited. In some cases, a re-evaluation of alternative disposal methods may be warranted if potential impacts appear especially severe. In most cases, however, further evaluation will determine that benefits derived for the proposed resource recovery facility outweigh the potentially adverse impacts associated with a particular site. For example, while a facility centrally located in a wasteshed would afford benefits in terms of the conservation of transport vehicle fuel, a site acceptable in all respects except central location, even perhaps located outside the wasteshed, may remain desirable because of resource recovery benefits and the lack of feasible alternative solid waste disposal methods.

The following portions of this Chapter discuss environmental factors that should be examined in detail following the conclusion of the site screening process and the tentative selection of an acceptable site.

TRAFFIC

The public impact of MSW vehicles travelling to and from

a resource recovery facility is a factor often overlooked or given less attention than needed during the environmental evaluation process. A facility may discharge water pollutants to a surface water body or release pollutants to the atmosphere, but these impacts usually do not affect the average citizen as directly as traffic congestion in his neighborhood or near his work place. For this reason, potential traffic impacts associated with resource recovery facilities are an important environmental factor to be considered and evaluated.

It is not uncommon for resource recovery facilities capable of processing 1800 tons or more of solid waste per day to require during daily peaks one refuse transportation vehicle per five to ten minutes for delivery of waste to the facility. (For purposes of calculating the number of vehicles required, a refuse packer truck holds approximately 5 tons of MSW and a transfer vehicle or trailer holds about 18 tons). The actual frequency depends largely on the ratio of packer trucks to transport vehicles utilized and the storage capacity of the facility. By providing excess storage capacity (a minimum of three times the average daily tonnage throughput), greater flexibility in the frequency of deliveries can be attained. Thus, the highest frequency of truck traffic can be scheduled to occur during daylight hours on business days if impacts on commuter rush-hour traffic are to be minimized. Alternatively, if commuter traffic is adversely affected and the roadways used do not pass through or near residential neighborhoods, high frequency truck traffic can be scheduled to occur throughout the night, from after the evening rush-hour to just before the next morning rush-hour period.

Such delivery scheduling assumes that the MSW collection system employs enough transfer stations to provide minimum storage or buffering capacity, thereby eliminating the need to deliver MSW directly to the resource recovery facility. An adequate number of transfer stations in conjunction with excess storage capacity will eliminate the need to schedule truck deliveries on week-ends or holidays.

The site selected for establishment of a resource recovery facility should be near a major highway to allow easy site access. Utilization of secondary roadways or streets is undesirable because of noise impacts and traffic congestion on routes not designed for heavy truck use. Often, where a site is not located in close proximity to a major arterial, a new access road must be constructed and designed with adequate load-bearing support and moderate grade. A surprising quantity of fuel can be wasted over a short distance by loaded solid waste transport vehicles required to pull a steep grade.

An additional consideration associated with traffic impacts is the provision of adequate queuing distance. The facility should be oriented such that queuing is at all times confined to the immediate site. An example showing the calculation of minimum queuing distance was provided in Chapter 4. A situation in which loaded solid waste vehicles spill over onto off-site roadways represents a serious flaw in the layout of the plant.

WATER POLLUTION

General

Excluding domestic wastewater produced on-site by employees, which is generally discharged to a municipal sewage system, wastewater generated by resource recovery facilities can be divided into two categories: process wastewater and thermal cooling water. Process wastewater includes boiler blowdown water, scrubber water, residue quench overflow, off-gas vapor condensate, and refuse slurry "white water". Thermal cooling water includes both once-through non-contact cooling water and evaporative cooling tower blowdown.

Process Wastewater

Depending on the type of resource recovery technology employed, process wastewater will vary in quality. This is especially true with regard to organic material and suspended solids concentrations. Table 8-1 lists the type of process wastewaters produced by currently available technologies.

TABLE 8-1

PROCESS WASTEWATERS PRODUCED BY
RESOURCE RECOVERY TECHNOLOGIES

Incineration	RDF Systems		Pyrolysis Systems	
Waterwall	Eco-Fuel II	Wet pulping	Andco-Torrax	Purox
Scrubber water	Scrubber water	Refuse slurry "white water"	Scrubber water	Slag quench overflow
Boiler blowdown			Residue quench overflow	Water vapor condensate
Residue quench overflow				

Scrubber water is listed in the above table because some existing facilities utilize wet scrubbers for air pollution control. Waterwall systems usually employ electrostatic precipitators (ESP) as air pollution control devices. The Eco-Fuel II process is amenable to the use of cloth fabric-filters as control devices in place of wet scrubbers.

Process wastewaters are variable in terms of quality but contain pollutant concentrations high enough to prohibit their direct discharge to surface waters. Some contain such high concentrations of solids, organics, or in unusual cases heavy metals, that pretreatment is required prior to their discharge to municipal sewage systems. Because of a combination of pretreatment requirements and raw water costs, significant advances have been made in developing closed-loop wastewater recycling systems at resource recovery facilities. While some processes such as wet pulp RDF lend themselves conveniently to closed-loop water recycling because of the nature of the system, on-site treatment and recycling has been extended to boiler blowdown water, scrubber water, and even residue quench overflow. A recently developed residue quench system utilizes minimal quantities of water allowing virtually no overflow and provides just enough moisture to facilitate residue handling and disposal. As a result of advances in wastewater recycling, treated process wastewater discharges on surface waters often has a minimal environmental impacts.

Cooling Water

In contrast to process wastewater, cooling water (including evaporative cooling tower blowdown and boiler blowdown) is usually discharged to surface waters. All resource recovery facilities that employ condenser-turbine units for generation of electricity must utilize a cooling water system. The volume of once-through cooling water used at a waterwall incineration facility can approach 50 million gallons per day, (using a 1,500 ton per day facility as a typical example), with cooling water temperature at the point of discharge varying from 70 to 100°F. The temperature of the discharge depends on both the temperature of the ambient intake water and the capacity at which the facility is operated. Other than the thermal component, once-through cooling water is generally equal in quality to surface water at the intake structure. An exception occurs in cases where ambient water has high solids or mineral concentrations requiring the addition of de-mineralizers prior to being pumped through the condenser-turbine unit. However, the concentation of such substances is usually low enough to be of little concern in terms of adverse impacts on water quality or aquatic biota.

Evaporative cooling tower blowdown involves the discharge of a much smaller volume of wastewater than in once-through cooling systems and does not involve the thermal component. However, of concern with the discharge of this blowdown to surface waters is the fact that de-mineralizers and algicides are commonly added to the circulating cooling water and are present in the discharge. A typical evaporative cooling tower will concentrate solids and other non-evaporative pollutants, such as metals, present in the cooling water by a factor of about three. Thus, blowdown water can contain high concentrations of contaminants, depending on the quality of the raw water sources, sometimes requiring pre-treatment prior to discharge to surface waters or a municipal sewer system.

Prinicipal environmental concerns associated with the discharge of cooling water to surface waters involve potential impacts on aquatic biota. These impacts include the entrainment of organisms in the discharge thermal plume. The first step in conducting this evaluation involves a determination of aquatic species present in the area of the proposed discharge. The magnitude and extent of the discharge thermal plume in the proposed receiving waters should next be examined. This analysis should concentrate on the degree of mixing provided by the receiving waters based on seasonal and annual flow rates (for rivers) or circulation patterns (for lakes), seasonal ambient water temperatures including vertical and horizontal profiles, and an examination of bottom contours.

Several engineering design factors are included in calculating the mixing of thermal cooling water with receiving waters. These involve the design of the outfall diffuser, its depth in the water, and the discharge flow rate of the thermal wastewater. Usually, these factors can be manipulated to reduce potential entrainment or other adverse effects of the thermal discharge if such impacts appear particularly severe, for example, diffuser port diameters may be decreased and a greater number of ports provided to increase discharge velocities thereby increasing the size of the mixing zone to effect greater dilution. If certain seasonal periods, such as late spring, are determined to be especially critical in terms of the reproductive cycle of some aquatic species, facility maintenance may be scheduled for that period. However, because mid-winter or mid-summer periods may be identified as critical for some species, periods when demand for recovered energy would be at a maximum, such maintenance scheduling is not always feasible in terms of impact mitigation.

In order to determine potential impacts resulting from the discharge of evaporative cooling tower blowdown to surface waters, the same evaluation procedure as described above for thermal cooling water would be applied. Instead of the thermal component of discharged cooling water, inorganic and organic contaminants would be the parameters of concern. In many cases, impacts on water quality and/or aquatic biota would be severe enough to require pretreatment prior to discharge of the blowdown to either surface waters or a municipal sewer system.

A final environmental consideration generally associated with cooling water systems involves the intake structure used either on once-through cooling systems or to draw make-up water for facilities equipped with an evaporative cooling tower. The concern with these structures involves potential excessive impingement of aquatic biota on the intake screens. Intake structures should be designed with screens having slots no greater than 1.25 millimeters wide and, in the case of non-travelling screens, with some type of air or water backwash system to periodically remove debris accumulated on the screen surface. This latter feature maintains maximum effective screen surface area, thereby greatly contributing to lower intake water velocities. Travelling screens, through the movement of the screen itself, periodically shed such debris. Regardless of the screen design, intake velocities should not exceed 0.8 feet per second to minimize impingement potential.

AIR POLLUTION
General

The resource recovery process with the greatest potential for air pollution is incineration. While certain other resource recovery processes generate pollutants that are emitted to the atmosphere, the potential for adverse impacts on ambient air quality is not as great as that of incineration.

The incineration process generates exhaust gases including carbon monoxide (CO), hydrocarbons (HC), oxides of nitrogen (NO_x), sulfur dioxide (SO_2), and particulate matter. Because MSW is a heterogenous mixture, ideal and uniform combustion conditions are difficult to achieve, leaving a substantial potential for emission of these pollutants. However, under the proper combination of turbulence, temperature, and retention time, substantial reduction of pollutant gaseous and particulate matter emissions can be achieved.

Particulates

Particulate matter emission rates fluctuate and are dependent upon several parameters: waste feed size and other characteristics (ash and volatile matter content), the type of charging system, furnace design, temperature and air quantity provided. These variables are neither constant from facility to facility, nor invariant within a particular facility. Optimal use of overfire and underfire air are key parameters in limiting particulate matter emissions. Entrainment of particulate matter and low furnace temperatures can result from too much excess air. Controlling underfire air, combined with proper amounts of overfire air and grate motion-induced waste agitation, will minimize entrained fly ash and stack particulate matter emissions. The USEPA has developed emission factors for particulate matter based on total feed material charged, as shown in Table 8-2.

TABLE 8-2

UNCONTROLLED PARTICULATE MATTER EMISSIONS FOR RESOURCE RECOVERY FACILITIES

Average		Range	
g/kg	lb/ton	g/kg	lb/ton
14	29	6.9-23	14-45

Source: Reference 1.

Studies have indicated that substantial quantities of small particles in the respirable fraction (approximately 15 microns, where 10^6 microns equal one meter) are emitted from MSW incinerators. Various heavy metals and trace elements are included in this fraction. The mechanisms believed responsible for trace element fine particulate matter generation may be either high temperature vaporization/condensation of low melting point metals or elements dispersed on a molecular level within a combustible material. Elements known to have particulate distributions include sodium, cesium, chlorine, bromine, copper, zinc, arsenic, silver, cadmium, indium, tin, antimony, and lead. Elements with a large particle predominance include calcium, aluminum, titanium, scandium, and lanthanum, originating, for the most part, from soils. The USEPA has provided a range for values of trace elements contained in particulate emissions from waterwall incineration resource recovery facilities. These are listed in Table 8-3.

TABLE 8-3

CONCENTRATIONS OF TRACE ELEMENTS IN
PARTICULATE MATTER EMISSIONS

Element	Concentrations (ppm)	
	Uncontrolled	Controlled
Antimony	260 - 620	460 - 1,000
Arsenic	50 - 70	50 - 100
Barium	270 - 540	270 - 540
Bromine	420 - 2,400	350 - 1,200
Cadmium	380 - 820	670 - 1,150
Chlorine	>10,000	>10,000
Chromium	50 - 560	130 - 260
Cobalt	10 - 100	5 - 50
Copper	420 - 590	620 - 800
Iron	970 - 1,090	2,000 - 2,130
Lead	11,600 - 17,500	18,100 - 34,200
Manganese	420 - 1,400	140 - 490
Nickel	[a]	[a]
Selenium	<90	<30
Silver	110 - 200	50 - 110
Tin	2,600 - 5,000	1,400 - 5,000
Zinc	>10,000	>10,000

[a] Data not available
Source: Reference 1.

Other Criteria Pollutants[1]

Sulfur dioxide emissions from incineration facilities show a wide range of values but are typically between 1.5 and

---

[1] Criteria pollutants are those for which National Ambient Air Quality Standards (NAAQS) have been promulgated (CO, non-methane hydrocarbons, NO_x, SO_2, lead and particulates).

3.0 pounds of pollutant per ton of MSW charged. Emissions of NO_x are typically low due to the relatively low temperatures at which MSW incinerator occurs. Hydrocarbon and carbon monoxide emissions are variable, depending on combustion techniques, operating practices, and combustibility of the MSW. The USEPA has recently published emission factors for some uncontrolled criteria pollutants emitted from waterwall incineration resource recovery facilities. These are shown in Table 8-4.

TABLE 8-4

UNCONTROLLED EMISSION FACTORS FOR MAJOR EMITTED
CRITERIA POLLUTANTS

	Emission Factors			
	Average		Range	
Pollutant	g/kg	lb/ton	g/kg	lb/ton
Sulfur oxides (as SO_2)	1.0	2.0	0.11- 3.2	0.21- 6.4
Nitrogen oxides (as NO_2)	0.8	1.6	0.46- 1.2	0.92- 2.3
Hydrocarbons (as CH_4)	0.06	0.12	0.013-0.12	0.027-0.24

Hazardous Air Pollutants

Recently, concerns have been raised about the emission of potentially hazardous air pollutants from resource recovery facilities, including heavy metals as discussed earlier. Other substances of concern include dioxins, polychlorinated biphenyls (PCB), and polycyclic aromatic hydrocarbons such as benzo(a)pyrene . Because these compounds are considered hazardous at very low concentrations, an analysis of potential impacts is very difficult to conduct since the heterogeneous nature of MSW cannot discount the possibility of trace amounts being present. High combustion temperatures (in excess of 2,000° Fahrenheit) in conjunction with minimal combustion gas retention times (at least 1.5 seconds) results in over 90 percent reduction in concentration of virtually any organic compound that may be present in the flue gas stream.

MSW combustion flue gases typically contain corrosive hydrogen chloride, a result of burning polyvinyl chloride or other chlorinated plastics found in the waste stream. The emission of HCl to the atmosphere is variable and dependent upon several factors: HCl absorption by alkaline bottom ashes, wet scrubber removal of substantial amounts of soluble HCl gas, and fly ash removal of absorbed HCl. Because of these influences, it is difficult to generalize a value of anticipated HCl emissions.

Table 8-5 is a compilation of incinerator emissions downstream of an electrostatic precipitator from over 15 industrial plants.

TABLE 8-5

EMISSION FACTORS FOR REFUSE BURNING
DOWNSTREAM OF AN ELECTROSTATIC PRECIPITATOR

Pollutant	Concentration PPM(a)	Concentration gr/dscf(b)	Pounds/Ton
Particulates		0.02	0.34
Sulfur Dioxide	80		2.4
Nitrogen Oxides at NO_2	75		1.6
Carbon Monoxide	150		1.9
Hydrocarbons	16		0.12
Hydrochloric Acid	200		3.4
Fluorides	6.5		0.06
Lead		0.00068	0.012
Mercury		0.00024	0.0064
Beryllium		0.000000003	0.000000051
Sulfuric Acid Mist		0.0023	0.04
Tetrachlorodibenzo-p-dioxins			0.00000001
Polynuclear Aromatics			0.00001
Polychlorinated Biphenyls			0.00013
Asbestos		not detected	
Hydrogen Sulfide		not detected	
Vinyl Chloride		not detected	
Reduced Sulfur		not detected	

(a) Parts volume per million parts volume.
(b) Corrected to 12% CO_2

Reference (2)

Control Technologies

At present, particulate matter is the only pollutant for which control devices have been installed at incineration facilities. Pollutants such as SO_2 and NO_x are not typically controlled except to the extent to which incinerator operation limits their generation and release. The potential also exists for emissions of carbon monoxide and hydrocarbons, however, the magnitude of these emissions is dependent on the heterogeneity of the solid waste charged, the charging mode (continuous or batch), and the extent to which proper operation is maintained and combustion zone parameters (air supply and temperatures) are regulated.

Particulate matter control technologies generally include baghouses (fabric filtration), wet scrubbers, and electrostatic precipitators. All three devices can achieve high removal efficiencies; however, each system has its inherent advantages and deficiencies. As such, the selection of a particular control device will depend on capital, operation, maintenance costs, and its design removal efficiency.

Baghouse Filters - Baghouses, as the name implies, are structures containing filter bags through which flue gases pass for the removal of particulate matter. These bags are made of a variety of natural or synthetic fabrics, the choice dependent on the temperature and chemical constituents of the combustion gas and the physical characteristics of the particulate matter emissions produced. A typical baghouse filter is shown in Figure 8-1.

Historically, baghouse use has been recommended for particulate matter control for many types of industries. These devices can collect particles as small as 0.4 micron. The principal advantages of baghouse systems include:

- o consistently high particle removal efficiencies over a wide spectrum of micron and sub-micron sized particles;

- o variations in inlet dust loadings or flue gas flow rates do not significantly effect removal efficiency for routinely cleaned filters;

- o corrosion and icing/rusting of equipment and wastewater treatment is avoided;

- o simplified operation and maintenance requirements; and

- o flexibility of design and configuration to suit installation requirements.

Deficiencies include:

- o potential for fabric blinding from adhesion and accretion of hygroscopic material and/or tarry subtances with resultant high pressure drops and large energy consumption;

- o shortened fabric life in the presence of acid or alkaline particles or gases at elevated temperatures;

- o potential for fire and explosion if oxidizable dust is collected; and

- o possible corrosion of structure due to acid gas condensation.

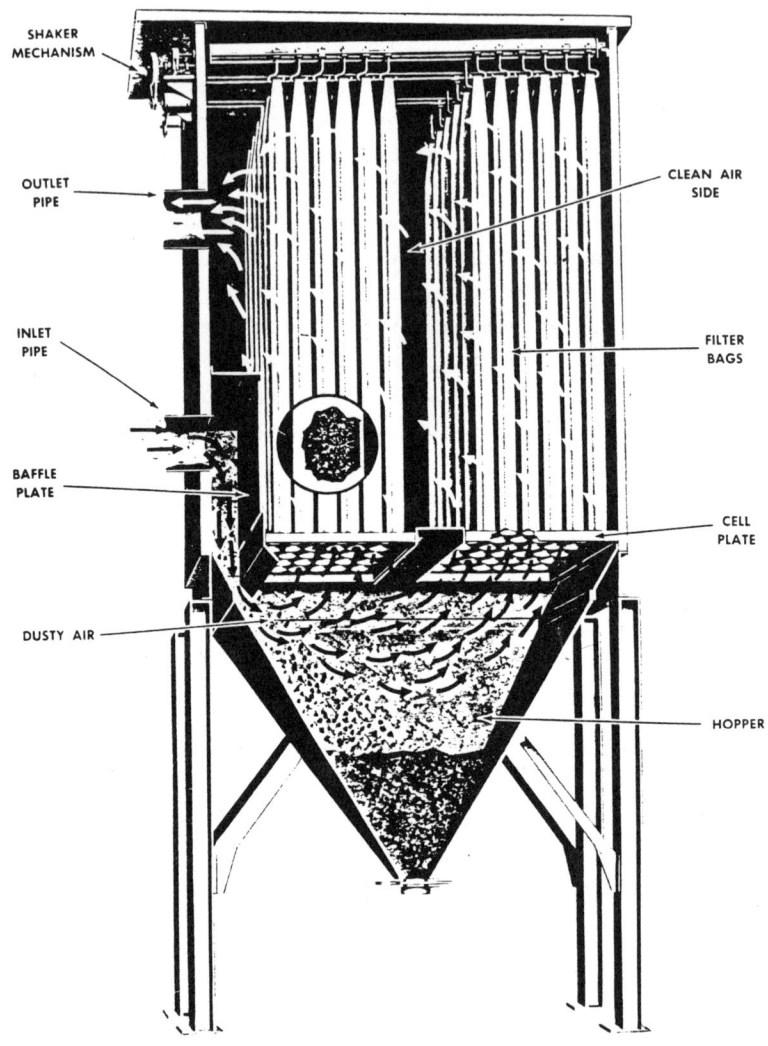

Figure 8-1
Typical baghouse filter with
mechanical shaking
(Wheelabrator Corp.)

Scrubbers - A scrubber is a device that employs a liquid to separate undesired particulates and gases from the flue gas flow although dry scrubbers, using a limited quantity of liquid injection, are used in industrial applications. Scrubbing liquids vary, the absorbing material selection dependent on product availability and desired scrubbing characteristics. Typically, low energy particulate matter collection devices are used in series with more effective equipment to minimize scouring. Two common types of scrubbers are centrifugal spray and venturi scrubbers, shown in Figures 8-2 and 8-3. The latter type requires considerable energy input, but can collect particles as small as 0.5 micron with 97 to 99 percent removal efficiency. Typical advantages of high energy wet scrubber use include:

- o simultaneous removal of particulate matter and acid and other components of exhaust gases;
- o particulate matter removal efficiency can be varied according to specified energy input;
- o high temperature flue gas streams can be accommodated;
- o dew point/moisture content of the incinerator exhaust gas will not affect equipment performance; and
- o operating conditions and choice of scrubbing medium can be varied to increase removal efficiencies.

Disadvantages include:

- o high energy input requirement to remove small particle fractions;
- o a cold water supply must be available;
- o high maintenance costs: equipment corrosion and erosion is unavoidable; and
- o wastewater treatment requirements.

<u>Electrostatic Precipitators</u> - Electrostatic precipitators are devices that capture particles by electrostatic attraction. They consist of small diameter negative electrodes (to increase the voltage potential) and a grounded positive plate. The negative electrodes impart an electric charge to the particles entrained in the gas flow which are attracted to the positive plate where they are collected and neutralized. Rappers hit the plates on a timed basis to discharge collected particulate. Some designs, wet precipitators, utilize water sprays to wash particulate from the plates. Figure 8-4 shows a typical electrostatic precipitator. Advantages include:

- o minimal pressure drop resulting in low power requirements;
- o relative insensitivity to high effluent gas temperatures;

218   Energy and Resource Recovery from Waste

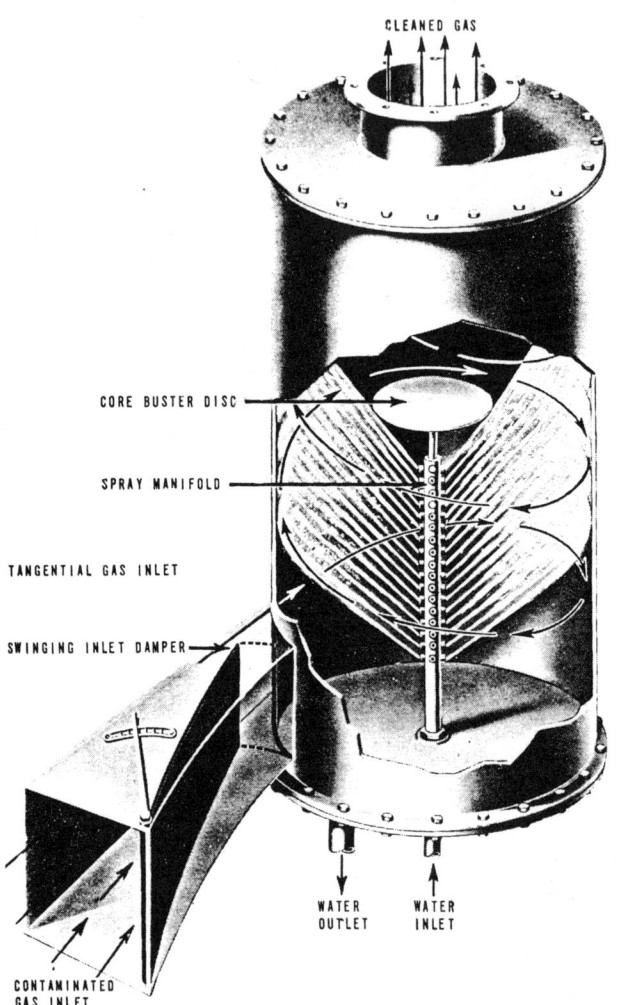

Figure 8-2
Cyclone Scrubber
(Chemical Construction Co.)

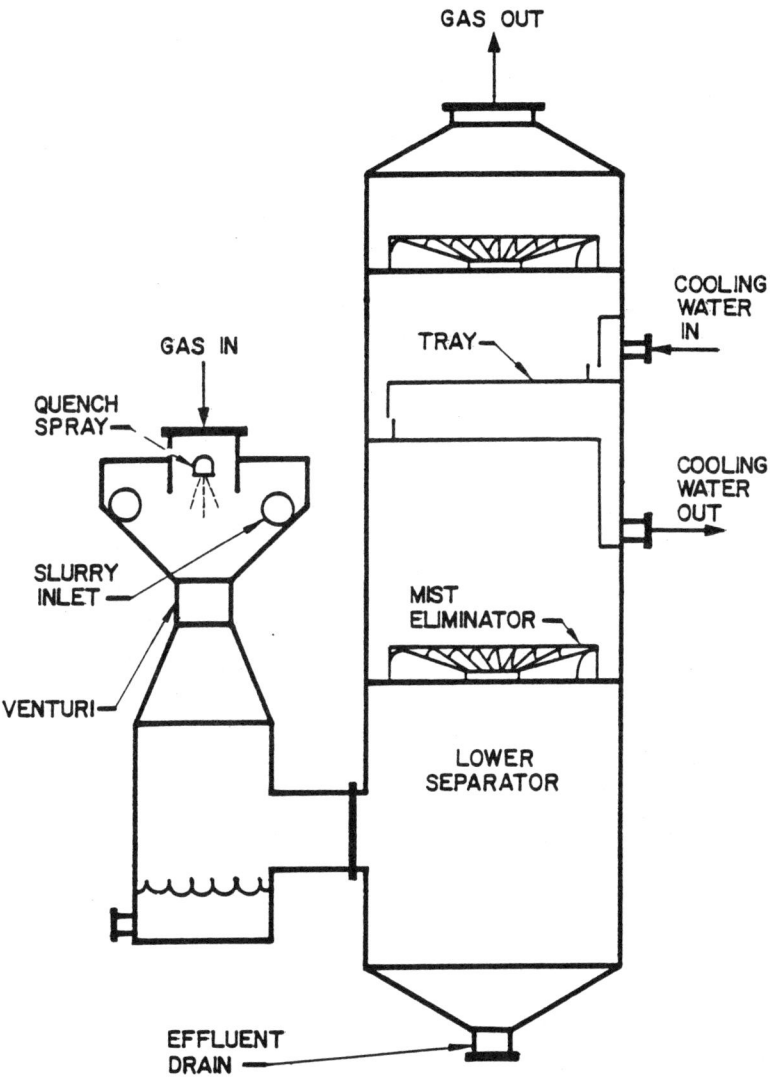

VENTURI SCRUBBER AND TRAY COOLER

Figure 8-3

220 Energy and Resource Recovery from Waste

o no wastewater treatment requirements (except for wet electrostatic precipitators).

Disadvantages include:

o high capital costs and large space requirements;

o sensitivity of the device to chemical composition of flue gas and differential particle resistivities;

o equipment is required for removal of collected soot;

o corrosion at low and high temperatures can occur due to acid gas condensation and acid gas attack in an oxidative atmosphere; and

o extensive electrical equipment is required.

Figure 8-4
Typical precipitator
(Joy Manufacturing Co.)

Control Technology Selection - Selecting an air pollution control device to meet particulate matter emission standards involves three factors:
- o consistency in removing micron and submicron sized particulates at high efficiencies;
- o ability to perform as required under various charging and operating conditions; and
- o minimal down-time.

There is no general solution for controlling particulate matter. Of the three primary types of control devices described, difficulties have been encountered with each system in either equipment maintenance or meeting existing particulate emission standards. Facilities employing fabric filtration have had a history of bag and baghouse corrosion and high opacity, although compliance with emission standards has been achieved. The use of scrubbers can result in difficulties in system maintenance and operation. Electrostatic precipitator control technology has shown a demonstrated capability to meet emissions standards although some test data have indicated performance deterioration with time. (This may be the result of a change in the type of refuse charged, poor operation and maintenance of the furnace, aging of the control device, or a combination of these factors.)

The evaluation of potential environmental impacts associated with the emission of air pollutants from waterwall incineration resource recovery facilities should include, at a minimum, the following considerations:
- o facility design parameters including effective stack height, exit gas temperatures, and exit gas velocities;
- o annual and seasonal meteorological factors such as wind speed and direction, air temperature, turbulence and mixing rates, and the influence of topographical features to channel or direct wind flow;
- o regional and local air quality including seasonal and annual ambient concentrations of air pollutants, especially particulate matter, sulfur dioxide, lead, and carbon monoxide; and
- o the emission rate of criteria pollutants (particulates, sulfur, dioxide, lead, nitrogen oxides, carbon monoxide, and non-methane hydrocarbons) estimated for the facility, especially under maximum charge conditions.

Because existing air pollution regulations, especially the USEPA Prevention of Significant Deterioration or PSD permit program, mandate specific procedures for conducting an environmental evaluation incorporating the above listed considerations, the methodology employed to conduct the evaluation will be discussed in detail in the final portion of this Chapter. Basically, the evaluation focuses on compliance with ambient air quality standards and the degree of air quality deterioration that may result. These standards were developed to protect both human health (primary standards) and wildlife, vegetation, structures, and aesthetic values (secondary standards).

AESTHETICS
General

Potential aesthetic impacts associated with resource recovery facilities include the generation of odors and noise in addition to the visual presence of an industrial-type facility. In general, these impacts do not vary in relation to the type of resource recovery technology employed since most odors and noise result from the transport, storage, and handling of MSW prior to its being processed.

Odor

The fact that MSW is odiferous cannot be ignored. Organic malodorous compounds generated by the decomposition of putrescible material can be detected at very low concentrations in ambient air. While both the threshold of detectability for these compounds and the level at which they would be classified a nuisance varies from individual to individual, it is generally agreed that odors released to the atmosphere should not be continuously discernible at any point beyond the facility property boundary by the average citizen. The term "continuously discernible" is used because the detection of odors beyond the property boundary is often directly related to air temperature, humidity, wind direction, and wind speed. Thus, even the best control program cannot guarantee that odors will be prevented from reaching areas beyond the facility property boundary at all times. For this reason, the most effective mitigation of odor impact involves proper siting of a resource recovery facility. Sites adjacent to residential neighborhoods or to particularly sensitive facilities such as churches, schools, hospitals, and nursing homes should be avoided when possible. Most odors associated with MSW will not be detected at a distance of more than 1,000 feet even under the worst combination of wind direction, humidity, and air temperature. Where location of a facility at this minimum distance to residential areas is impossible, measures can be taken to reduce odor impacts. These include the maintenance of negative pressure in structures housing tipping floor areas, storage pits, shredding equipment, residue handling and sorting equipment, and residue storage areas. By far the most significant sources of odor are the tipping floor, storage pits, and shredders. The most practical method of maintaining negative pressures is by drawing ventilation air from these structures for combustion. For technologies employing no combustion, or starved-air combustion processes, a charcoal filter system may have to be installed in these structures with internal air being drawn through it prior to release to the outside atmosphere.

Odors from MSW transport vehicles can be minimized by keeping the refuse storage compartment closed or covered during transit. The same is true for residue transport vehicles removing material from the facility for final disposal.

Because of high combustion temperatures associated with MSW incineration, malodorous organic compounds are generally destroyed resulting in minimal odor from the release of combustion gases. An exception to this occurs when such facilities are not continuously operated or when they are brought back into operation following maintenance. In these cases, low combustion temperatures during start-up and shut-down do not result in the destruction of malodorous organics. Proper operation dictates that these facilities be brought up in temperature with supplemental fuel before MSW or RDF is admitted.

Noise

In general, noise control procedures are similar to the odor control techniques discussed above. The enclosure of tipping floor areas, shredders, and residue handling areas in structures greatly attenuates noise. As a general rule, noise levels measured at the facility property boundary should not exceed 45 to 60 dBA depending on the characteristics of the surrounding neighborhood. Proper maintenance of equipment, including MSW transport vehicles, is an important aspect of noise control.

Visual

The best way to effectively mitigate visual impacts associated with resource recovery facilities is through proper siting and selection of an access route for vehicles. Appropriate landscaping and the taking advantage of on-site topographic features may assist in making the facility less conspicuous from ground-level but it is usually impossible to screen a 100-foot tall boiler housing structure, for example, or a 200-foot tall stack, from view. Consideration of land use in the surrounding area of the proposed site is by far the best approach to potential visual impacts.

RESIDUE DISPOSAL

Unmarketable residues produced by MSW resource recovery facilities include bottom ash from waterwall incineration furnaces, slag and char from pyrolysis systems, and ESP and baghouse fly ash. Facilities using a wet scrubber for control of air emissions, and which use lagoons for treatment of scrubber wastewater, require periodic dredging of ash material for final disposal. Shredded refuse derived fuel (RDF) facilities do not employ processes that generate thermally reduced residues. However, the heavy fraction produced during the classification step of the shredded RDF process contains putrescibles and unrecyclable material from the MSW and must be transported off-site for final disposal. Some RDF processes utilize heavy organics from air classification as a fuel for process heat. Incineration of this fraction reduces residue disposal problems.

Slag and char produced by pyrolysis systems are relatively inert materials free of organic material. This residue is suitable for use as concrete aggregate and "glassphalt" although markets for such use do not exist in many areas. As a result, this residue must often be disposed of by conventional landfilling.

Bottom ash from waterwall incineration and fly ash from an ESP or baghouse are also suitable construction aggregate materials. However, as is the case with slag and char from pyrolysis, markets for use of this material as aggregate often do not exist in close proximity to the resource recovery facility. Therefore, it must be disposed of through landfill operations.

Ash residue from incineration is basically inert although putrescible content is generally low (less than five percent by weight). Combustibles usually range from one to four percent by weight, and moisture content is generally 20 percent or lower. Because of the presence of soluble metals in this residue, and the presence of putrescible matter, albeit at low levels, concerns with utilizing conventional landfilling methods to dispose of this material focus on potential ground-water pollution from leachate. For this reason, elutriate or leachate tests may be required to determine the concentration and types of metals present in the residue. This analysis will determine the need for impervious liners, ground-water monitoring wells, and/or leachate collection systems to protect surface and ground-water quality. As part of the site selection process for location of an ash disposal landfill facility, analyses of soil materials from the proposed site should be conducted to determine permeability, grain size, cation exchange capacity, and pH. Borings should also be taken to determine depth to bedrock and depth to ground water. The results of these analyses represent input to the decision on necessity of impervious liners and monitoring wells.

While MSW resource recovery technologies greatly reduce the volume of waste requiring landfilling, the selection of a suitable site for residue disposal should take long-term capacity into consideration. Depending on the resource recovery process employed, residue may represent form 10 to 40 percent by weight of the charged MSW and from 5 to 25 percent by volume of the MSW. Also, depending on the process, bulk density of residue ranges from 60 to 100 pounds per cubic foot. A significant factor in determining the quantity of residue requiring landfilling is the degree of which metals will be reclaimed either before the MSW is processed or from the residue. Available markets for sale as construction aggregate is another important factor in calculating quantities of residue for landfill disposal. The selected landfill site should provide enough capacity for an operating life of at least 10 to 15 years.

PERMITS AND REGULATORY REQUIREMENTS
General

Resource recovery facilities must satisfy environmental regulatory requirements and obtain construction and operating permits imposed by federal, state and often local levels of government. An important benefit derived from careful and comprehensive environmental evaluations conducted throughout facility planning, especially with regard to siting, is that of greatly facilitating the preparation of required permit applications and supporting documentation. Early identification of potential environmental problems allows the development of reasonable alternative facility construction and/or operating plans before substantial time and capital has been invested in a particular process or mode of operation.

NEPA

Requirements of the National Environmental Policy Act of 1969 (NEPA) apply to resource recovery projects in which federal funding has been obtained. In this case, either an Environmental Impact Statement (EIS) must be prepared or a declaration of Finding of No Significant Impact issued by the funding agency. Some states have adopted requirements similar to NEPA that apply to major actions as determined by established criteria. These criteria usually address total acreage involved in the project, the sensitivity of the sites or site affected by the project, and the size of the facility based on average or maximum tonnage capacity. Basically, an EIS is an information document prepared for review by regulatory agencies and the public. It sets forth the reasoning or logic used in development of a proposed action and environmental implications associated with implementing the action. Finally, it serves to balance, in a concise yet comprehensive way, negative versus positive aspects and impacts of the action to arrive at a conclusion concerning its overall acceptability. If public input is solicited during facility planning, the EIS can focus on those issues of greatest local concern and thereby expedite public review and acceptance of the final document.

Environmental Considerations   227

Air Quality Regulations

By far the most complex and time-consuming permit action associated with resource recovery facilities involves air quality. Table 8-6 summarizes the federal air quality regulations which apply to a resource recovery facility. These regulations are discussed below.

New Source Performance Standards - USEPA New Source Performance Standards (NSPS), 40 CFR Part 60, Subpart E, regulate particulate matter emissions from municipal incinerators and resource recovery facilities built after August 17, 1971 having a design capacity exceeding 50 tons/day and burning more than 50 percent solid waste. Emissions cannot contain particulate matter in excess of 0.08 grains/dscf corrected to 12 percent CO_2. No other emissions are regulated by NSPS. Performance tests are required to demonstrate compliance with the NSPS.

Typically, states regulate particulate matter emissions and opacity, the level of strigency being dependent on unit capacity and age, and the type of waste chrged. All 50 states have such standards, with some developing regulations specific to the control of trace metal emissions. As a rule, particulate matter standards are based on the weight of the waste charged or volume of the flue gas with either a fixed emission limitation irrespective of charging capacity or variable standards as a function of charging capacity.

For a facility charging less than 50 tons per day of waste material, the applicant would have to comply only with state air pollution permitting procedures. This process involves meeting state emission and opacity regulations and other requirements that may include opacity monitoring, control of odor, and hazardous pollutant emissions, specification of state-of-the-art control technologies, and other system monitoring and controls.

A resource recovery facility designed to charge more than 50 tons/day, but less than 250 tons/day of municipal solid waste must comply with Federal NSPS as well as state emission standards.

If the proposed facility will charge more than 250 tons/day licensing may involve passing through the USEPA Prevention of Significant Deterioration (PSD) discussed below.

TABLE 8-6

AIR QUALITY REGULATIONS AFFECTING RESOURCE RECOVERY FACILITIES

Facility Size	State Emission Standards	Federal NSPS	Federal PSD Program	Facility Location	Control Technology	Allowable Impact on Ambient Air
< 50 tpd	Do apply	Does not apply	Does not apply	Has no effect	As required to meet state emissions standards	As specified by State
> 50 tpd <250 tpd	Do apply	Particulates: 0.08 gr/dscf	Does not apply	Has no effect	As required to meet state emission standards and/or NSPS	As specified by State
>250 tpd	Do apply	Particulates: 0.08 gr/dscf	Does apply if PE of criteria pollutant >100 tpy	Attainment area	BACT	May not exceed NAQS for criteria pollutants. May not exceed allowable increment for SO_x and particulates. Non-criteria pollutants handled on a case-by-case basis.
				Nonattainment area	LAER	No degradation allowed; must make good faith effort to obtain off-set for any emissions.

Criteria Pollutants - Table 8-7 lists pollutants for which National Ambient Air Quality Standards (NAAQS) have been established.

These pollutants are the focus of regulatory review and must be addressed in terms of emission rates and impact on ambient concentrations in the permit process.

TABLE 8-7

SUMMARY OF FEDERAL AMBIENT AIR STANDARDS

Pollutant	Interval*	Federal Std (Primary)	
		PPM	$\mu g/m^3$
Sulfur Dioxide	(Annual) Arithmetic mean	0.03	80
	24-hr conc	0.14	365
	3-hr conc	--	--
	1-hr conc	0.50	1300
Particulate (suspended)	(Annual) Geometric mean	75	--
	24-hr conc		
Carbon Monoxide	8-hr conc	9	10 mg/m^3
	1-hr conc	35	40 mg/m^3
Hydrocarbons (non-methane)	3hr conc (6-9 am)	0.24	160
Nitrogen Dioxide	(Annual) Arithmetric mean	0.05	100
Lead	Calendar quarter	--	1.5

* Except for annual values, the Federal standards are values not to be exceeded more than once per year unless otherwise noted.

For particulates and SO_x, allowable increments (degree of air quality deterioration allowed) under the PSD program have been specified. These increments are listed in Table 8-8.

These increments apply in attainment areas (an attainment area is one in which NAAQS criterial are met for a particular pollutant) only. In non-attainment areas the PSD emission offset requirements apply (see PSD subsection below).

In addition to meeting NSPS requirements, emissions from resource recovery facilities must not violate the allowable increment.

Increments have not been established as yet for other criteria pollutants. For these other pollutants a resource recovery facility can not result in the ambient air qualify being a violation of NAAQS.

TABLE 8-8

PERMISSIBLE INCREMENTS ABOVE BASELINE CONCENTRATIONS ACCORDING TO FEDERAL PREVENTION OF SIGNIFICANT DETERIORATION (PSD) REGULATIONS

Pollutant	Maximum Allowable Increase ($\mu g/m^3$)
Particulate Matter	
Annual geometric mean	19
24-hour maximum	37
Sulfur Dioxide	
Annual arithmetic mean	20
24-hour maximum	91
3-hour maximum	512

Note: The increments presented above apply to Class II Regions under the PSD regulations.

Non-Criteria Pollutants - Pollutants for which NAAQS have not been established are termed non-criteria pollutants. Examples include heavy metals (other than lead), chlorides, and chlorinated hydrocarbons (dioxin, PCB, DDT and others). These pollutants are regulated under the federal PSD program and in states which have established emission limitations and/or ambient air standards for specific pollutants. Each of these pollutants must be addressed on a case by case basis with the applicant presenting data to demonstrate that the emission level of that pollutant from the proposed facility will have no significant impact.

PSD Program - The PSD permitting process is triggered by the size of the facility (capable of charging more than 250 tons/day MSW) and the potential to emit (more than 100 tons/year of a criteria pollutant).

For facilities under the PSD program, located in an attainment area, allowable PSD increments and NAAQS may not be exceeded.

For PSD regulated facilities located in non-attainment areas, emissions must not contribute to further air quality degradation. Since any combustion facility must emit some pollutants, it is required to obtain offsets for such emissions. Offsets are defined as concommitant reduction in emission in the non-attainment area.

The only exception to this requirement is for a resource recovery facility which must make a good faith effort to find offsets, but is not bound by the requirement. Under the PSD program two levels of pollutant control technology may be required. In attainment areas the applicant must employ Best Available Control Technology (BACT), which is proven, available control technology. Consideration of costs and benefits are allowable in selecting BACT.

In non-attainment areas a higher level of emission control, called the Lowest Achievable Emission Rate (LAER) is required. LAER is the best existing technology that can be applied to the facility regardless of cost or lack of a proven operating record.

The PSD Application - If PSD is triggered, demonstration studies must be conducted as part of the permit application. They include:

- o Best Available Control Technology (BACT) Assessments.
- o Ambient Air Quality Monitoring.
- o Source Impact Modeling.
- o Additional Impact Analyses.

Construction and operation of a large resource recovery facility has many complex and related impacts, including economics, energy penalties, and environmental impacts. The BACT analysis addresses these issues and further determines the control strategy necessary for the particular pollutant. The primary purpose of BACT is to minimize PSD increment (maximum allowable source impact concentration increases over background air quality) consumption, and can thus be considered the most important aspect of the PSD process. When considering BACT, energy and economic costs of emission controls should be reasonable while considering direct and residual risks to the environment. As such, a BACT assessment depends on site-specific factors and is conducted on a case by case basis.

Before a PSD permit can be granted, the operator of a resource recovery facilities must demonstrate that neither a NAAQS nor an allowable increment will be exceeded as a result of the emissions from the facility. An air quality analysis consists of ambient air quality monitoring and source impact modeling, and must be conducted for each regulated pollutant subject to PSD review that will be emitted by the resource recovery facility.

A crucial, time consuming aspect of the PSD process is the requirement for the operator to monitor for up to one year prior to construction, using approved methods and techniques, all regulated pollutants emitted in significant quantities. The monitoring requirement for a particular pollutant can be waived if predicted source impact values are below certain levels.

A diffusion (source impact) model is a quantitative description and linking of pollutant sources, transport mechanisms, interactions, and the resulting effect on the environment. The diffusion model, given the proper source and meteorological data, will estimate total air quality: baseline air quality (from monitoring data) plus increment consumption. To obtain a PSD permit, the applicant must demonstrate that the emissions from the resource recovery facility will not cause or contribute to violations of total suspended particulate matter or SO_2 increments.

All applicants preparing a PSD submittal are required to conduct additional impact analyses for each pollutant subject to review. These analyses consist of:

- o Associated growth impacts.
- o Impacts on soils and vegetation.
- o Visibility impairment (where applicable).

Water Quality Regulations

Compliance with the National Pollutant Discharge Elimination System (NPDES) program is much less complex in comparison to the PSD program. Many states have assumed responsibility for this regulatory program designed to protect surface waters from point sources of water pollution. As discussed earlier wastewater recycling has been developed for some resource recovery facilities rendering compliance with NPDES much easier since no process wastewater discharge is involved. However, the discharge of thermal cooling water requires a NPDES or state equivalent permit under the program. In order to obtain this permit, the applicant must demonstrate that the thermal discharge will not affect biota in the receiving waters to any significant extent nor offset designated use of the waters for recreational or commercial purposes. Although the final determination of the permitting agency is usally based on existing water quality data from the receiving waters, discharge temperature of the cooling water, and a review of aquatic biota known to inhabit the water, additional data gathering by the applicant may be required. This may include collecting ambient water temperatures and generating horizontal and vertical profiles and/or may include computer modeling to determine the degree of mixing and the dimensions of the mixing zone.

If cooling water intake structures and/or discharge diffuser pipes are to be located in navigable waters or in waterways having a flow rate greater than five cubic feet per second, a US Army Corps of Engineers permit will be required. Also, if dredging to facilitate construction of these structures or filling of adjacent waterways for any reason is proposed, the Corps will review the proposed dredging and filling as part of the permit application process. The applicant must provide detailed design information relative to intake strucutres, discharge structures, and the exact location of each. For dredge and/or fill operations, the applicant must discuss potential impacts on aquatic biota, the dredging operation to be employed, and the nature and characteritics of the proposed fill material. Before this permit can be processed, the applicant must request a water quality determination from the State in which the action will take place. This water quality certification is provided to the Corps for review of acceptable conditions in the surface waters in which the permitted action would take place. A public hearing concerning the Corps action on the application may or may not be required.

Construction and Operating Permits

Construction and operating permits for a MSW processing facilities, including resource recovery facilities, are required by all states. This requirement covers both the resource recovery facility and the final disposal methods relative to solid residue. To obtain these permits, an applicant must provide detailed engineering design information on the facility, including measures to control vectors, odors and noise. Information relative to the capacity of the proposed facility must be provided in terms of average and maximum daily throughput and annual average capacity. If a landfill is proposed for disposal of solid residue, the application must include information on soil characteristics, depth to ground-water, depth to bedrock, ground-water quality, quality of the residue to be disposed of especially in terms of metals and potentially hazardous substances, location and proximity of surface water bodies, and capacity and operating conditions of the proposed facility. A public hearing usually is required prior to the issuance of a final permit to construct and operate.

REFERENCES

1. Rinaldi, G.M. et.al., Monsanto Research Corporation, An Evaluation of Emission Factors for Waste-to-Energy Systems: Execution Summary, Office of Research and Development, USEPA, July, 1979.

2. O'Connell, Stotler, G., Clark, R., Emissions and Emission Control in Modern Municipal Incinerators, Proceedings of the 1982 ASME National Waste Processing Conference.

9
Institutional Factors

INTRODUCTION

The technical, environmental, and economic factors which influence resource recovery feasibility and implementation have been discussed in other chapters. Equally important are institutional factors which include:
- o solid waste flow control;
- o procurement methodology;
- o implementation alternatives;
- o legal constraints;
- o financing alternatives; and
- o funding assistance.

SOLID WASTE FLOW CONTROL
General

As discussed in Chapter 2, an accurate prediction of the solid waste quantities to be received by the resource recovery facility is critical to estimating facility economics and evaluating feasibility. However, correct estimates of solid waste generation and collection do not in themselves insure that the solid waste collected will, in fact, be delivered to the resource recovery facility. Such assurance is vital to the feasibility of the resource recovery program.

In some cases, the implementing entity is a municipality which itself is responsible for solid waste collection within its jurisdiction. A decision by such a municipality to implement a resource recovery project then automatically commits the solid waste under the direct control of the municipality to that project. However, solid waste collected by private carters, or other participating municipalities, require more formal commitment. Mechanisms for obtaining the commitment include:

- contracts
- flow control ordinances
- artificially low or zero tipping fees

Contract

The most direct method of controlling the solid waste flow is to contract directly with the municipality or private carter collecting the waste. Typically such contracts would:
- specify a tipping fee and an escalation mechanism;
- commit the resource recovery facility to accept up to a predetermined waste quantity;
- commit the collector of solid waste to deliver at least a minimum solid waste quantity; and
- establish a mechanism for sharing revenues and revenue increases from recovered energy and material products.

Examples of the contractual method of flow control include CEA's Bridgeport, CT facility, and Saugus, MA.

The contractual method of flow control requires that the resource recovery implementation entity assume substantial risks with regard to project economics, since no municipality or private carter will commit itself to a long-term contract without a firm price specified in advance. Since contracts must be negotiated and signed before the facility is designed and built, the risks of construction or operating cost overruns, or delays in start-up, fall to the implementing entity.

Flow Control Laws

These laws permit a city, county, solid waste authority, or other governmental entity to direct municipal and private solid waste collectors to use specified facilities for disposal, or control of refuse flow. Examples of this method include Monroe County, NY; Akron, OH; and the State of New Jersey.

This method of flow control has worked in some cases, but questions exist as to its legality and economic efficiency.

In several instances the legality of flow control has been challenged, usually by affected private carters, on grounds of restraint of trade, interference with interstate commerce, and the taking of private property. One recent noteworthy case of this type was Glenwillow Landfill vs. Akron, where, in December 1979, the US District Court upheld the right of the City to direct private carters to use the Akron recycle energy system. This case is now on appeal. In other instances, municipalities have been unwilling to enforce flow control laws, due to concern that they will not withstand legal challenge.

Another argument against the use of laws to control solid waste flow is that this approach eliminates landfill as competition for resource recovery, thus eliminating an important incentive for efficient operation.

Artificially Low Tipping Fees

A simple method of flow control is to set the tipping fee at the resource recovery facility below the prevailing rate for alternative disposal options. This effectively eliminates all objections to using the facility, but often requires that the facility be subsidized by some other revenue source, typically ad valorem taxation.

One way of implementing this approach is through the establishment of a solid waste disposal district or authority which finances the resource recovery facility, in part, through property taxes levied against everyone within the facility's service area. These taxes are then used to lower the tipping fee to the point where there is no incentive to avoid using the facility.

This method is effective, and has the advantage of permitting "fine-tuning" of the solid waste flow to the resource recovery facility by adjustments to the tipping fee. However, it can often be difficult to implement due to public opposition to property tax increases in general.

PROCUREMENT METHODOLOGY
General

There are several reasons why the procurement of resource

recovery systems and services is more complex than that of traditional public works projects:

- o Recource recovery is new in the United States, and considerable uncertainty exists with regard to its reliability and cost.

- o Resource recovery projects are often considered as business operations than as traditional public works projects, and are expected to be financially self-sufficient, as a business operation would be.

- o Resource recovery projects are often multi-jurisdictional, requiring solid waste from a number of jursdictions.

- o Resource recovery projects are often the subject of intense political pressures because of the public's strong interest in their siting and cost.

- o Resource recovery projects require long-term contracts, sometimes up to 20 years, for operation, waste supply, and sale of recovered materials and energy products.

Procurement Alternatives

There are two basic procurement methods for resource recovery:

- o Formal bids; or
- o Negotiated Procurement.

Formal Bids - The traditional methods by which a governmental body procures goods or services is to receive formal bids, and to award the contract or purchase order to the lowest bidder. This procurement method, required by law in many jurisdictions, is intended to prevent collusion in the award of governmental contracts, and to insure low price.

Two procurements are actually required in this method: first the municipality selects an architect-engineer who developes detailed plans and specifications for the proposed project. These contract documents then form the basis for the second procurement, solicitation of bids prepared by contractors to construct and/or operate the proposed facility.

Legal constraints designed to prevent collusion permit only very limited exchanges of information and modifications to specifications during the bidding process. For this reason, the contract documents must be extremely detailed and complete, and their preparation can be expensive and time consuming.

In this method, contract award is required by law to be given to the lowest responsive bidder.

In summary, a formal bid procurement generally involves the following steps:

- o Selection of an architect-engineer
- o Preparation of detailed engineering plans and specifications
- o Advertising for bids
- o Public opening and recording of bids
- o Evaluation of bids for exceptions, responsiveness, and mistakes
- o Identification of the low bidder
- o Judgment on capability to perform
- o Award

<u>Negotiated Procurement</u> - Negotiated procurement is often used to obtain complex services and non-standard items. Its main advantage is that it permits a dialogue to develop between the bidder and the municipality. The requirements of the municipality can be modified to reflect the capabilities of the bidders and the costs involved. In addition, it offers the opportunity to consider non-quantifiable factors, such as facility aesthetics or bidder management expertise.

A negotiated procurement usually involves a Request for Proposal (RFP) soliciation. An RFP is a comprehensive document which describes what is required of proposers with regard to facility capacity, siting, materials and energy recovery, and so forth. An RFP does not generally detail the way in which these results are to be achieved, but rather sets performance standards which the proposers must meet.

An RFP or non-negotiated performance generally includes the following steps:

- o Preparation of an RFP
- o Advertisement for proposals
- o Receipt and evaluation of proposals
- o Negotiate with bidders
- o Contractor selection
- o Contract negotiation
- o Contract award

IMPLEMENTATION ALTERNATIVES

General

There are five basic ways to implement a resource recovery program. These are:

- Conventional
- Turnkey
- Full service
- Full service with public ownership
- Modified full service

Figure 9-1 summarizes the various roles and responsibilities in each approach.

ACQUISITION APPROACH	PROJECT ELEMENT						PROC METHOD	EXAMPLE
	DESIGN	CONSTRUCT	SUPV CONST	SHAKE DOWN	OPERATE	OWN		
CONVENTIONAL	PRIVATE	GOVT ────────────────────────────────▶					A&E DESIGN NON NEGOTIATED	AKRON OHIO
TURNKEY	PRIVATE ──────────────────▶				GOVT ─────▶		NEGOTIATED	TACOMA WASH.
FULL SERVICE	PRIVATE ─────────────────────────────▶						NEGOTIATED	HEMPSTEAD N.Y.
FULL SERVICE WITH GOVT OWNERSHIP	PRIVATE ───────────────────────▶					GOVT	NEGOTIATED	BRIDGEPORT CONN.
MODIFIED FULL SERVICE	PRIVATE	GOVT	PRIVATE ─────────────▶			GOVT	* NEGOTIATED	MONROE COUNTY N.Y.

* EXCEPT CONSTRUCTION ELEMENT

Figure 9-1
Implementation Approaches
(Reference 4)

Conventional

This is the traditional approach to public works projects. The municipality hires an architect-engineer who prepares contract documents which are the basis for formal bids (see Procurement subsection). The facility is then constructed by a contractor, is owned by the municipality, and operated by municipal employees.

This method is perhaps the lowest cost method, because of its reliance on competitive bids, but it exposes the municipality to the greatest risk. Because design engineers do not generally guarantee the performance of their designs, and because the responsibility for design, construction, and

Institutional Factors 241

operation is divided, if the facility fails to perform as expected the responsibility will likely fall on the municipality.

Turnkey

In this approach the responsibilities for design, construction and startup are assigned to one contractor. This reduces the risk to the municipality substantially, since if the facility does not meet performance specifications it need not be accepted. In addition, design and construction cost overruns are borne by the contractor.

However, the municipality does own and operate the facility, and the risks of long-term operation and marketing arrangements are still its responsibility.

Full Service

In the full service approach a private contractor designs, constructs, owns, and operates the resource recovery facility. In effect, the municipality procures a service (solid waste disposal) rather than a facility.

This approach shifts almost all the risk to the full-service contractor, but at considerable expense: the full service contractor normally charges a substantial premium for assuming this risk.

Full Service with Public Ownership

In this variation of the full-service approach, the facility is owned by the public rather than private sector. The contractor is still responsible for facility design, construction, and operation. This approach is actually an extension of the turnkey approach in that the contractor is responsible for long-term operation of the facility, as well as design, construction, and startup. It offers the potential for lower costs than the full-service approach, since the contractors service charge would not include a return on equity. In order for this approach to be viable, however, the public sector must have sufficient confidence in the technology proposed to be willing to risk ownership.

Modified Full Service

This is a hybrid approach combining private sector

design, construction supervision startup and operation with public sector ownership and conventional low bid contractor construction. This approach is useful when a full service approach is desired, but competitive bidding laws must be met.

LEGAL CONSTRAINTS

General

Resource recovery is a new undertaken for many municipalities, and in some cases legal barriers exist which may prevent utilization of a preferred or cost-effective approach. These barriers are generally not intended to disadvantage resource recovery, but rather are traditional limitations on public sector activities which simply are not appropriate when the public sector is acting in an entrepreneurial role.

Legal constraints may affect resource recovery in the following areas:

- solid waste flow control;
- competitive bidding requirements;
- long-term contracts; and
- regulation of the sale of recovered energy products.

Solid Waste Flow Control

This issue has been discussed earlier in this Chapter. Briefly, the question is whether or not a municipality may require, by ordinance, that solid waste be delivered to specified disposal facilities. This question is now being litigated in State and Federal courts.

Competitive Bidding Requirements

In some jurisdictions it is a requirement of law that the public sector procure goods and services by competitive bidding, with award to the lowest responsive bidder. This requirement precludes all negotiated RFP-type resource recovery procurement.

Most jurisdictions exempt professional services and other specialized services from competitive bidding requirements, and in some cases it has been argued successfully that procuring resource recovery services falls within these categories. In other cases, it has been necessary to obtain a specific exemption from bidding requirements for resource recovery.

Another area in which bidding laws affect resource recovery is in the area of sale of recovered energy and materials products. In some jurisdictions a municipality must sell any surplus commodities by annual bid. This requirement is obviously inappropriate to resource recovery, where long-term contracts are desirable.

Long-Term Contracts

Resource recovery projects typically involve the investment of tens or hundreds of millions of dollars: such investments are not possible without long-term contracts for solid waste supply, sale of recovered energy and materials products, and operation of the facility. Often 10 years is the minimum feasible duration of such contracts, and 15 to 20 years is preferable.

The ability of a municipality to contract for long terms is generally specified by state law, and often contracts less than 20 years are required. This issue should be carefully checked during resource recovery feasibility evaluation.

Regulation of the Sale of Recovered Energy Products

The sale of electric power is a regulated monopoly in the United States, and in many states the sale of steam and hot water are also regulated. In many cases public utilities have exclusive franchise areas within which competition is not permitted. Generally, a critical issue is whether the resource recovery facility is to market energy to the public in general, which is usually regulated, or only within the governmental unit which owns or operates the facility.

One way of avoiding this issue is for the local electric power utility to own the turbine-generator, and for the resource recovery facility to sell steam to the utility. If the turbine is within the resource recovery site, such sale of steam is generally not regulated.

PURPA

One goal of the Public Utilities Regulatory Policies Act of 1978 (PURPA) is to encourage cogeneration and small power production. Cogeneration refers to equipment that generates both electricity and useful thermal energy (i.e. heat or steam). In an effort to implement this goal, the Federal

Energy Regulatory Commission (FERC) has adopted regulations that provide the following:

- o Utilities must purchase electricity from qualifying cogeneration and small power production facilities at an appropriate rate (to be determined by state regulatory agencies, but incorporating cost elements of capital offset and fuel replacement).

- o Qualifying facilities that produce and sell electric energy, are exempt from certain federal and state regulations pertaining to electric utilities.

- o Utilities must provide qualifying facilities with electric energy and other certain types of services which may be requested by a qualifying facility to supplement or back up those facilities own generation.

- o Implementation of the PURPA regulations is the responsibility of state regulatory authorities and non-regulated electric utilities.

Qualifying Facilities

Qualifying facilities fall into two categories: small power production facilities and cogeneration facilities. To qualify as a small power production facility a producer must meet the following criteria:

- o Power production capacity must be less than 80 megawatts. (However, to qualify for an exemption from the Federal Power Act a small power production facility's capacity must be less than 30 megawatts).

- o Biomass, waste, renewable resources or any combination thereof must account for greater than 50 percent of the facilities total energy source. Oil, natural gas, and coal in aggregate may not exceed 25 percent of the total energy input of the facility during a year.

- o Not more than 50 percent of the equity interest in the facility may be owned by an electrid utility, utilities, or subsidiaries thereof.

To qualify as a cogeneration facility a producer must meet the following criteria:

- o The same ownership criteria as small power production facilities.

- o It is not a new (installed after 3/13/80) diesel cogeneration facility.

- o If it is a topping cycle facility (first produces electric power, then thermal energy) it must meet the following operating and efficiency standards:

Institutional Factors

Operating Standard -

1. Useful thermal energy (heat or steam) must equal 5 percent or more of the total energy output.

Efficiency Standards -

1. If natural gas or oil are not used for energy input and installation began prior to 3/13/81, then there is no efficiency standard.

2. If natural gas or oil are used for energy input and installation began on or after 3/13/80 then the following efficiency standards apply: If thermal energy output is less than 15 percent of the total energy output then the useful power output plus one half of the thermal output must be no less than 45 percent of the total energy input of natural gas and oil to the facility. If thermal energy output is equal to or greater than 15 percent of the total energy output then the useful power output plus one half of the thermal output must be no less than 42.5 percent of the total energy input of natural gas and oil to the facility.

o If it is a bottoming-cycle facility (first produces thermal energy, then electric power) it must meet the following efficiency standards:

1. If natural gas or oil are not used for any of the energy input as supplementary firing and installation began prior to 3/13/80, then there is no efficiency standard.

2. If natural gas or oil are used for any of the energy input as supplementary firing and installation began on or after 3/13/80, then the following efficiency standard applies. The power output of the facility during any calendar year must be no less than 45 percent of the energy input of natural gas and oil for supplementary firing.

An additional consideration, subject to section 292.205(c) of the PURPA regulations, is a cogeneration facility using natural gas may be exempted from incremental pricing.

Once a facility has determined that it is a qualifying facility, it is eligible for the benefits available to it through PURPA. These benefits originating from the electric utility are discussed below.

Under PURPA electric utilities are obligated to qualifying facilities for the following:

o Purchase any energy and capacity made available by a qualifying facility.

o Sell any energy and capacity requested by a qualifying facility.

o Interconnect with a qualifying facility when necessary to accomplish purchases or sales. This obligation is waived if, solely by reason of the purchase or sales over the interconnect, the electric utility would become subject to regulations as a public utility under Part II of the Federal Power Act.

Interconnection costs are to be paid by the qualifying facility and are to be set at a level comparable to the interconnection costs of other customers with similar load characteristics. Rates for purchases and sales are determined through a complex set of rules. Briefly, these rules require the utility to pay for energy from qualifying facilities based on "avoided costs" or the money the utility saves by not having to generate the energy provided by the qualifying facility. The avoided costs of a utility are usually its highest costs, since the last generating equipment to be brought on-line is the utility's most expensive equipment.

The rules also discuss the utility paying for "avoided capacity costs". The primary issue here is the determination of how much of the qualifying facility's power capability is considered reliable enough to displace the need for new power plants. (In Florida, the Public Service Commission recently adopted rules deeming a qualifying facility as eligible for a capacity credit if it provides energy to a utility at 70 percent equivalent availability). Finally, rates for sales are to be comparable to rates for sales to other customers served by the utility.

Implementation of the PURPA regulations is the responsibility of the appropriate state regulatory authority and nonregulated electric utilities. As of March 20, 1981, the state and nonregulated electric utilities were required to have implemented the PURPA regulations. There are a number of challenges to provision of PURPA in federal courts regarding constitutionality and other issues. Some states have implemented major portions of this statute while others are awaiting the outcome of pending litigation.

FINANCING ALTERNATIVES

General

The selection of the financing alternative should be based on the economic benefits the affected parties can derive from the financing arrangements and the risks that must be assumed to receive those benefits. The financing alternatives available for resource recovery projects can be organized according to public versus private financial instruments and project versus general obligation financing. Table 9-1 illustrates how financial alternatives can be categorized. This certainly is not the only way to categorize these alternatives and some of these mechanisms could fall in other categories depending on the contracting liabilities the participating parties assume.

TABLE 9-1

FINANCIAL OPTIONS

	Public Financial Instruments	Private Financial Instruments
Project Financing	Municipal and Industrial Revenue Bonds	Leveraged Leasing Pollution Control Revenue Bonds (without pledge of corporate general obligation)
General Obligation Financing	Municipal General Obligations	Pollution Control Revenue Bonds (with pledge of corporate general obligation)

Reference 6.

Project Financing Alternatives

Project financing guarantees the bond principal and the interest repayment with the expected project revenues, which are recovered product sales and tipping fees. Under this type of financing these revenues must offset all future recovery and operating/maintenance costs.

Municipal and industrial revenue bonds, pollution control revenue bonds and leveraged leasing are examples of project financing mechanisms.

Municipal revenue bonds are long-term, tax exempt obligations that are payable solely from the project revenues or revenues from other facilites owned by the public body. Municipalities issue the bonds, which are also used to finance such services as sewers, bridges, and housing projects.

Industrial revenue bonds are long-term, tax exempt bonds that can be issued by a public benefit corporation to promote industrial or economic development, which in some cases includes resource recovery projects. When this type of financial mechanism is utilized the project being financed is leased or the bond proceeds are loaned to a private corporation.

Pollution control revenue bonds (PCRB) are similar to industrial revenue bonds. They are long-term, tax exempt bonds that may be issued by a public agency to, in effect, obtain low-cost financing for a private enterprise. However, there are several differences. PCRBs can be used only to finance pollution control equipment and they may be considered as project financing or general obligation financing, depending on whether the private corporation involved must provide a corporate guarantee for the debt service payments. If this pledge is required then the PCRB would be a general obligation type of financing. The utilization of PCRB's has been very limited in resource recovery projects.

Leveraged leasing is a financing alternative, although it is technically not a financial instrument. It involves several financial mechanisms that combine to provide the project participants with either tax benefits or lower financing costs. These benefits are achieved by the leveraging of an equity investor's funds and the pass-through to the user/lessee (via lower lease rentals) of the tax benefits that are derived from owning the project. This type of financing should be employed when the user/lessee cannot fully utilize the tax benefits of ownership. If the user/lessee can use the tax benefits it may be difficult to justify leverage leasing for the project.

General Obligation Financing Alternatives

General obligation financing is backed by the full faith and credit of the issuing municipality. This means that the credit rating of the issuing public body is the determinant of the cost of financing (interest rate) and not the technical or marketing risk associated with a particular project.

General obligation (GO) bonds are the primary instrument used in general obligation financing. Other examples of general obligation financing are corporate bonds and PCRBs with a pledge of corporate general obligation, both of which are rarely, if ever, used to finance resource recovery projects. GO bonds require voter approval and tend to carry the lowest interest rate of any financial alternative. This low interest rate is a result of the minimal risk assumed by investors because of the guarantees the municipality supplies. In determining if GO bonding is the preferred alternative factors that deserve consideration include: debt-ceiling on the public bodies' borrowing capacity, municipalities current credit rating, impact of the financing on the municipality's credit rating, and voter resistance to the bond offering.

There are a number of factors which must be considered in evaluation of the alternative financing methods.[3] These factors include:

- The terms of the proposed debt structure (interest rate and repayment schedule);

- The availability of tax benefits to private owners;

- The availability and cost of equity capital;

- The availability and restrictions on the use of federal assistance programs (loan guarantess, price support loans, construction grants etc);

- The availability of state funds, and the restrictions on the projects' financing by using such funds; and

- The costs associated with developing and implementing a financing program (underwriting costs, legal fees, insurance, taxes etc).

Evaluation of the finacing alternatives and selection of the optimal financial package must be incorporated into the entire development of the project. In addition, the needs, goals, and objectives of project participants should be met by the financing program and it should effectively distribute the economic benefits and associated risks to the appropriate party.

Example 9-1 illustrates the calculation of project economics, assuming revenue bond financing.

EXAMPLE 9-1

Consider a resource recovery facility receiving 432,000 tpy with the following estimated costs and revenues:

Facility Capital Cost:	$ 70,000,000
O&M:	4,000,000/yr
Steam Revenues:	9,000,000/yr
Electric Power Revenues:	3,000,000/yr
Ferrous Revenues:	200,000/yr

Assume the use of industrial revenue bonds at 10 percent, 20 years, with a coverage ratio of 1.5. Legal and miscellaneous costs can be assumed to be $250,000 and the cost for financing will be approximately 2-1/2% of the bond issued. Calculate breakeven tipping fee. (The capital recovery factor can be found from Table 9-2).

1. Compute project capital cost.

Facility capital cost =	$70,000,000
Engineering @ 8%	5,600,000
Subtotal	$75,600,000
Contingencies @ 10%	7,560,000
Start-up @ 3%	2,270,000
Bond Issue & Builders Risk Insurance @ 1%	756,000
Project Capital Cost =	$86,186,000

2. Compute bond issue size

   Bond issue size (S) is the facility capital cost, plus one year's debt service, plus interest during construction, plus 90 days working capital, plus finance costs, plus legal and other costs.

   a. Project capital cost = $86,186,000
   b. *Debt service = S x CRF$^{n=20}_{i=10\%}$ = 0.1175 S
   c. Interest during construction = (construction period/2) (i) (s) = 3/2 x 0.10 x S = 0.1500 S
   d. 90 days working capital
      = $\frac{O\&M}{4}$ = $\frac{4,000,000}{4}$ = 1,000,000
   e. Finance costs = 0.025 S
   f. Legal & other costs = 250,000
   g. S = sum of a. through f.
      S = 87,436,000 + 0.2925 S
      S = 123,584,000

3. Calculate debt service:

   Debt service = Bond issue x CRF
   = 123,584,000 x 0.1175 = $ 14,521,000

4. Calculate annual revenues:

Steam revenues	=	$ 9,000,000
Electric power revenue	=	3,000,000
Ferrous revenues	=	200,000

*See Table 9-2 for CRF

	Interest on bond reserve, debt service x interest		
	14,521,000 x 0.10	=	1,452,000
	Annual revenues	=	$ 13,652,000

5. Compute annual cost:

Debt service	=	$ 14,521,000
O&M	=	4,000,000
Annual revenues		(13,652,000)
Net annual cost		$ 4,869,000

6. Compute breakeven tipping fee:

Funds required for debt service 1.5 x debt service 1.5 x 14,251,000	=	$ 21,377,000
Funds available for debt service Revenues less O&M $13,652,000 - $4,000,000	=	(9,652,000)
Required tipping fee revenues	=	$ 11,725,000
Required tipping fee for 432,000 tpy	=	$27.14/ton

7. Compute retained revenue:

Tipping fee revenues	=	$ 11,725,000
Net annual cost	=	4,869,000
Retained revenues	=	$ 16,594,000

TABLE 9-2
CAPITAL RECOVERY FACTORS
Interest - Percent

Years	7	8	9	10	11	12	13	14	15	16
5	0.2439	0.2505	0.2571	0.2638	0.2706	0.2774	0.2853	0.2913	0.2983	0.3054
6	0.2098	0.2163	0.2229	0.2296	0.2364	0.2432	0.2502	0.2572	0.2642	0.2714
7	0.1856	0.1921	0.1987	0.2054	0.2122	0.2191	0.2261	0.2332	0.2404	0.2476
8	0.1675	0.1740	0.1807	0.1874	0.1943	0.2013	0.2084	0.2156	0.2229	0.2302
9	0.1535	0.1601	0.1668	0.1736	0.1806	0.1877	0.1949	0.2022	0.2096	0.2171
10	0.1424	0.1490	0.1558	0.1627	0.1698	0.1770	0.1843	0.1917	0.1993	0.2069
11	0.1334	0.1401	0.1469	0.1540	0.1611	0.1684	0.1758	0.1834	0.1911	0.1989
12	0.1259	0.1327	0.1397	0.1468	0.1540	0.1614	0.1690	0.1767	0.1845	0.1924
13	0.1997	0.1265	0.1336	0.1408	0.1482	0.1557	0.1634	0.1712	0.1791	0.1872
14	0.1443	0.1213	0.1284	0.1357	0.1432	0.1509	0.1587	0.1666	0.1747	0.1829
15	0.1098	0.1168	0.1241	0.1315	0.1391	0.1468	0.1547	0.1628	0.1710	0.1794
16	0.1059	0.1130	0.1205	0.1278	0.1355	0.1434	0.1514	0.1596	0.1679	0.1764
17	0.1024	0.1096	0.1770	0.1247	0.1325	0.1405	0.1486	0.1569	0.1654	0.1740
18	0.0994	0.1067	0.1142	0.1219	0.1298	0.1379	0.1462	0.1546	0.1632	0.1719
19	0.0968	0.1041	0.1117	0.1195	0.1267	0.1358	0.1441	0.1527	0.1613	0.1701
20	0.0944	0.1019	0.1095	0.1175	0.1256	0.1339	0.1424	0.1510	0.1598	0.1687
21	0.0923	0.0998	0.1076	0.1156	0.1238	0.1332	0.1408	0.1495	0.1584	0.1674
22	0.0904	0.0980	0.1059	0.1140	0.1223	0.1308	0.1395	0.1483	0.1573	0.1664
23	0.0887	0.0964	0.1044	0.1126	0.1210	0.1296	0.1383	0.1472	0.1563	0.1654
24	0.0872	0.0950	0.1030	0.1113	0.1198	0.1285	0.1373	0.1463	0.1554	0.1647
25	0.0858	0.0937	0.1018	0.1102	0.1187	0.1275	0.1364	0.1455	0.1547	0.1640
26	0.0846	0.0925	0.1007	0.1092	0.1178	0.1267	0.1357	0.1448	0.1541	0.1634
27	0.0834	0.0914	0.0997	0.1083	0.1170	0.1259	0.1350	0.1442	0.1535	0.1630
28	0.0824	0.0905	0.0989	0.1075	0.1163	0.1252	0.1344	0.1437	0.1531	0.1625
29	0.0814	0.0896	0.0981	0.1067	0.1156	0.1247	0.1339	0.1432	0.1527	0.1622
30	0.0806	0.0888	0.0973	0.1061	0.1150	0.1241	0.1334	0.1428	0.1523	0.1619

FUNDING ASSISTANCE
General

In the past, federal encouragement for investment in resource recovery has been lagging. However, recent legislation have substantially enhanced the availability of financial incentives for investment in resource recovery. A number of federal programs provide resource recovery projects with significant indirect assistance, primarily in the form of tax benefits. Included in this group of incentives are:

- Investment and Energy Tax Credits

    These make available, for private owners, a 10 percent investment tax credit on new facilities and a 10 percent energy tax credit, which is proportionally less if the property is financed with tax-exempt industrial bonds.

- Industrial Development or Revenue Bonds

    This type of bonding permits a private organization or person to employ tax exempt bonding to finance a resource recovery project and thereby obtain a lower interest rate on the financing.

- Accelerated Depreciation

    This incentive allows private owners to accelerate the depreciation of a resource recovery facility for tax purposes. For facilities that qualify, the guidelines allow an 8 to 10 year life.

- Interest or Rental Cost Deduction

    Private owners are permitted a tax deduction for interest paid on debt or, for leased facilities, the entire rental payment.

Other programs offering financial incentives to resource recovery projects include the benefits available from the Public Utilities Regulatory Policy Act (PURPA), which was reviewed in this chapter, where electricity, as it pertains to energy markets, is discussed. In addition, the Department of Energy (DOE) has several programs it is administering under the Energy Security Act or "synfuels bill," are aimed at providing assistance to Municipal Waste to Energy Projects, which include resource recovery projects. These programs include construction loan guarantees, where the DOE guarantees the principal and interest on a loan to construct a facility, and price support loans, which are made available only under specific conditions, including that the energy produced at the facility is sold.

An effective financial program, sensitive to the needs of the participants and designed to expedite implementation of the project, should be the result of a thorough evaluation of the available financial incentives and alternatives.

REFERENCES

1. USEPA, *Procurement, Resource Recovery Plant Implementation: Guides for Municipal Officials*, SW-157.5, 1976.

2. Brown, Michael D., and Diana L. Powers, "Resource Recovery Systems, Part III: Implementation", *Solid Wastes Management* November 1980.

3. Aldrich, Robert H., and Rene Rofe, "Resource Recovery: An Investment Opportunity for the '80s" *NCRR Bulletin*, December 1980.

4. Mitre Corporation, *Resource Recovery Acquisition Strategies and Decisions*, (Conference Paper), May, 1975.

5. Bolton, R.E., *PURPA* (a private communication).

6. Randol, R.E., *Resource Recovery Plant Implementation Guide for Municipal Officials - Financing*, USEPA, 2nd Printing, (SW-157.4), 1977.

Appendix A

Market Assessment Documents

This appendix provides samples of some of the documents that should be developed in the course of performing a market assessment. As samples, these documents are presented with the purpose of demonstrating how such documents could be developed. Specific projects may have differing informational requirements and the marketing documents and tools should be tailored to satisfy these requirements.

The following is a listing of the documents contained in this appendix:

- o Exhibit A1 - Potential Materials User Questionnaire

- o Exhibit A2 - Potential Energy Products User Questionnaire

- o Exhibit A3 - Letter of Interest

- o Exhibit A4 - Advance Letter of Intent to Bid for the Purchase of Recovered Products

Exhibit A1

POTENTIAL MATERIALS USER QUESTIONNAIRE

From our initial information, it appears that you may be a potenital purchaser of a product(s) produced in the planned municipal solid waste resource recovery program. Please complete this questionnaire as comprehensively as possible, and return with any further information you feel is pertinent.

1. Ability to use product. What is the overall daily tonnage of the product that your facility(ies) could use? _____

Appendix A—Market Assessment Documents   255

2. What are the facility(ies) (names, locations) and the related daily product tonnages that would be involved?
    _____
    _____
    _____
    _____

3. What type of product transportation mode to your facility(ies) would your require? _____truck _____rail _____optional

4. Would you be willing to execute a five year _____ one year _____ longer term ( ) _____ contract for the purchase of the tonnage?

5. What type of pricing formula would you accept? Note one, or rank high to low preference.

    - Variable fee, according to an established market commodity comparison?
      If so, which specific commodity and comparison? Would there be a percentage of this commodity price below which the price would never go?

    - Fixed price for the entire length of the contract?

    - Fixed annual price, negotiated before each contract year?

6. What price would you offer as a floor price, below which the price of material delivered would never fall during the contract period?

7. Please attach a copy of any specifications you feel are initially desirable for the product.

8. What other considerations or requirements do you feel are necessary to mention at this time? _____
    _____
    _____

Exhibit A2

POTENTIAL ENERGY PRODUCTS USER QUESTIONNAIRE

Company Name _____
Address _____

Contact _____
Brief Description of Company Product(s): _____
_____
_____

Present Annual Fuel Consumption   Used for:

    Coal (Tons) _____   _____
    Oil (Barrels) _____   _____
    Gas (Million Cubic Feet) _____   _____
    Electricity (Kilowatts) _____   _____

Description of Generation Equipment
    No. of Boilers and Locations_____
_____

    Type Boilers and Their Manufacturers_____
_____

    Average Steam Demand (Pounds per Hour):_____

    Baseload Steam Demand (Pounds per Hour:_____

Steam Conditions: Temperature_____ Pressure_____

    Special Requirements?_____

Steam Use Pattern:_____ Hours/Day:_____Days/Week

    Would you require firm service?_____

    Do you plan shutdowns on a_____weekly;_____monthly;
    _____yearly basis.

    Is your operation subject to major fluctuations in steam usage:
    _____daily? _____seasonal?
    Scope and magnitude of fluctuations:_____

Future Energy Plans

    1. Do future plans call for installing new energy producing
    capability?
    _____Yes _____No
    If "yes", please describe:
_____
_____
_____

    2. Do future plans call for phasing out certain energy producing
    capability? _____Yes _____No If "yes", please describe:
_____
_____
_____

    3. Do future plans call for changing fuels used in present energy
    production equipment? _____Yes _____No
    If "yes", please describe:
_____
_____
_____

Interest in Derived Energy Products (See Attachment I for more detailed
                                        descriptions)

    At an attractive price, would you be interested in

    1. Solid Fuel Undensified (5500 BTU/lb.): _____Yes _____ No
    Quantify-_____ tons per_____.

    2. Solid Fuel Densified (5500 BTU/lb.): _____Yes _____No
    Quantify-_____ tons per_____.

3. Steam:
   _____°F _____ psig
   _____ lbs/hr. max.
   _____ lbs/hr. min
   _____ million lbs. per year.

4. Pyrolysis Gas:
   _____ million Cu Feet per year (150-299 BTU/scf and/or 300-400 BTU/scf.).

5. Methane Gas:
   _____ million Cu Feet per year (1000 BTU/scf).

6. Pyrolysis Oil: _____ Yes _____ No
   _____ barrels per year.

7. Electricity: _____ Yes _____ No
   _____ Kilowatts per year.

Cost
1. Present steam cost: $_____ 1,000 pounds.
2. Present fuel cost: $_____ per gal./ton/cu ft./million Btu.
3. At an attractive price, would you consider a long term contract for steam or fuel? _____
Comments: _____
_____
_____
_____

Exhibit A3
LETTER OF INTEREST

Dear _____:

The XYZ corporation with facilities located at 100 XYZ Street, Anywhere, USA is interested in negotiating an agreement to purchase _____ from the proposed resource recovery plant to be constructed at 5 XYZ Street, Anywhere, USA.

Our facility has been expanding over the past few years and future expansion plans could necessitate a need for up to _____ of _____ per day.

Of course, the terms of a negotiated agreement would have to be economically attractive to the XYZ Corporation.

Respectfully,

_____

Exhibit A4

ADVANCE LETTER OF INTENT TO BID FOR THE PURCHASE OF RECOVERED PRODUCTS*

Whereas, the _____ Corporation (hereinafter called the CORPORATION) endorses resource recovery from municipal solid waste as a means toward a cleaner environment and preservation of natural resources, and

Whereas, the CORPORATION recognizes the need to develop firm expressions of intent to purchase materials or energy products recovered from waste within known financial parameters as part of the planning process for a new endeavor such as this, and

Whereas,_____(hereinafter called the DEVELOPMENT AGENCY), is evaluating the prospects of substituting resource recovery for its traditional means of solid waste disposal, and

Whereas, the DEVELOPMENT AGENCY recognizes the need to establish financial data for the determination of the economic feasibility of processing up to_____ tons per day of _____ (hereinafter known as the PRODUCT) in a form usuable and acceptable to the CORPORATION according to the Specifications attached to this Agreement and made part hereof.

(a) It will be a firm bid for five (5) years offering an Exchange Price either fixed or related to a commodity quote, and if the Exchange Price is not fixed, it will offer a Floor Price which will not fall during the term of the contract.

(b-1) If the Exchange Price to be paid by the CORPORATION is to be a fixed dollar amount per unit of product, f.o.b. the recovery facility (or the CORPORATION'S plant - choose one), the bid shall not be less than _____ per ton.

(b-2) If the Exchange Price is to be based on a commodity quote, the monthly Exchange Price shall relate to the quotation at the close of that month for _____ (the same or the appropriate analogous commodity and location) as published in the last issue of the month of _____ (Fill in source of quote using the (mid-range or highside, or lowside choose one) of the quote, f.o.b. the recover facility (or the CORPORATION'S plant - choose one).
If the Exchange Price is to be bid in terms of a percentage of the quoted price, the Exchange Price shall not be bid at less than _____ percentage of appropriate quote as defined above.

(c) If the Exchange Price is not fixed, a Floor Price will be bid which will not be below $_____ per ton f.o.b. (fill in dollar amount) the recovery facility (or CORPORATION'S plant - choose one).

(d) The CORPORATION shall retain the right to reject any material delivered which does not meet Specifications. Such rejection will be at the expense of the resource recovery plant.

* Source: Resource Recovery Plant Implementation: Guides for Municipal Officials - Markets Yvonne M. Garbe and Steven J. Levy, U.S. Environmental Protection Agency, 1976.

Appendix A—Market Assessment Documents   259

(e) The bid will be subject to _force majeure_.

(f) It will be noted the Additional Conditions of the CORPORATIONS covering general terms and conditions of purchase, acceptance delivery, arbitration, weights, and downgrading not explicity covered in this Letter of Intent or by reference, will be negotiated according to good business practices and include such additional conditions as are attached to this Agreement and made a part hereof.

(g) This Advance Letter of Intent to bid is null and void if during the period between its execution and the actual bid or negotiated contract the CORPORATION'S plant ceases operation or not longer has a use for this or equivalent grade of recovered PRODUCT. The DEVELOPMENT AGENCY shall further recognize that a clause similar to this shall be incorporated in the actual bid when made or contract when signed.

(h) This Advance Letter of Intent may be assigned by the DEVELOPMENT AGENCY.

THEREFORE, in consideration of the fact that the legal authority to sell recovered products may rest upon a requirement to advertise for the purchase of such products, it is mutually agreed between the CORPORATION and the DEVELOPMENT AGENCY that:

I. The CORPORATION, as an expression of its support of the municipal solid waste recovery program, agrees to:

(1) offer herein a firm commitment to bid for the purchase of the recovered PRODUCT at prices not less than those entered here should the DEVELOPMENT AGENCY be required or decide to effect a competitive procurement, and

(2) agree that if public bidding is not necessary and not the course chosen by the DEVELOPMENT AGENCY then the conditions of this Letter of Intent may be considered as a bona fide offer to purchase the recovered PRODUCT at prices not less than those recovered here.

(3) respond should a bid be required with a bona fide offer to purchase which will include the following:

II. The DEVELOPMENT AGENCY agrees:

(1) to see that the recovery plant establishes specification assurance procedures for the recovered PRODUCT, using good industrial quality control practiced in recognition of the CORPORATION'S Use technology as practiced in their _____ plant, so as to produce and offer the recovered PRODUCT for sale in a form and to the required Specification, useable in the plant with minimum alterations to present processing technology and business practices, and

(2) to require, should a contract be effected as a result of the Advance Letter of Intent, that the PRODUCT be delivered to the CORPORATION according to conditions and prices determined herein and not diverted to a spot market which may on occasion be higher than the Exchange Price determined by the pricing relationship set forth here or as modified by the Contract.

(3) that should the CORPORATION'S plant, as specified herein, become saturated in its ability to handle the recovered PRODUCT as a result of other Letters of Intent issued by the CORPORATION being converted into firm contracts for delivery and purchase prior to effecting such arrangements as a result of this commitment, the provisions of this Advance Letter of Intent become null and void.

The CORPORATION will communicate to the DEVELOPMENT AGENCY that information about is use technology and business practices which the CORPORATION at its sole discretion shall consider necessary so as to assure receipt of the recovered material in form and cleanliness necessary for use by the CORPORATION. Such communication shall be on a nonconfidential basis, unless otherwise subject to a confidentiality agreement.

This Advance Letter of Intent shall become null and void on _____ unless effected into a contractual relationship or mutually extended by both the CORPORATION and DEVELOPMENT AGENCY.

Witnessed by: DEVELOPMENT AGENCY

_____   By:_____
_____   CORPORATION

Witnessed by:

_____   _____
_____   _____

Appendix B
Recovered Materials Specifications

This appendix provides samples of specifications that may be utilized to evaluate some of the materials that can be recovered from MSW. The following is a listing of the materials contained in this appendix.

- o   Exhibit B1 - Paper Stock Institute - Partial listing of Paper Specifications

- o   Exhibit B2 - Target Specification for Recovered Newsprint

- o   Exhibit B3 - Target Specification for Recovered Old Corrugated Boxes

- o   Exhibit B4 - Chemical Analyses for Aluminum Scrap, Maximum Percent by Weight

- o   Exhibit B5 - Standard Specification for Waste Glass as a Raw Material for the Manufacture of Glass Containers

Exhibit B1
PAPER STOCK INSTITUTE -
PARTIAL LISTING OF PAPER SPECIFICATIONS

#1 NEWS

Consists of newspaper packed in bales of not less than 54 inches in length, containing less than 5 percent of other papers. Prohibitive materials may not exceed . . . . .1/2 of 1%
Total Outthrows* may not exceed 2%

* Outthrows are contaminants which make the product unsuitable for comsumption at the grade specified. Outthrows usually consist of materials which are compatible with the paper-making process. Examples include cloth bindings, chipboard, string bindings and glassine.

SUPER NEWS

Consists of sorted fresh newspapers, not sunburned, packed in bales of not less than 60 inches in length, free from papers other than news and containing not more than the normal percentage of rotogravure and colored sections.
Prohibitive materials . . . . None permitted
Total Outthrows may not exceed. 2%

SPECIAL NEWS DE-INK QUALITY

Consists of sorted, fresh, dry newspapers, not sunburned. Packed in bales not less than 60 inches in length, free from magazines, white blank, pressroom over-issues, and paper other than news, and containing not more than the normal percentage of rotogravure and colored sections.
This packing must be free from tar.
Prohibitive materials . . . . None permitted
Total Outthrows . . . . 1/4 of 1%

OVER-ISSUE NEWS

Consists of unused over-run regular newspapers printed on newsprint, baled or securely tied in bundles, and shall contain not more than the normal percentage of rotogravure and colored sections.
Prohibitive materials . . . . None permitted
Total Outthrows . . . . None permitted

Source: Paper Stock Standards and Practices Circular,
          Paper Stock Institute of America, August, 1972.

Exhibit B2

TARGET SPECIFICATION FOR RECOVERED NEWSPRINT

Grade Title:	Recovered News (equivalent grades--Folded News, Regular News, Ordinary Folded News, No. 1 News)
Description:	Consists of folded newspaper including the normal percentage of rotogravure and colored sections
Packing:	Packed in bales of standard dimensions, not less than 54 inches long, approximately 1,000 to 1,500 pounds per bale.
Moisture:	Packed air dry
Prohibitive Materials:	Less than 1/2 percent
Outthrows*:	Less than 2 percent

* Outthrows are contaminants which make the product unsuitable for comsumption at the grade specified. Outthrows usually consist of materials which are compatible with the paper-making process. Examples include cloth bindings, chipboard, string bindings and glassine.

Appendix B—Recovered Materials Specifications 263

Water Solubles:	Less than 2 percent of the acceptable paper (Note a)
Organic Solubles:	Less than 2 percent of the acceptable paper (Note b)
Ash:	Less than 1 percent of the acceptable paper (Note c)

Note a: Determined by ASTM D-1162 or equivalent
Note b: Determined by ASTM D-1804 or equivalent
Note c: Determined by ASTM D- 586 or equivalent

Source: Alter, H. and Reeves, W.R. Specifications for Materials Recovered from Municipal Refuse, The National Center for Resource Recovery, May 1975, NTIS #PB-242 540.

Exhibit B3

TARGET SPECIFICATION FOR RECOVERED OLD CORRUGATED BOXES

Grade Title:	Recovered Old Corrugated Boxes
Description:	Consists of used corrugated containers having liners of jute or kraft
Packing:	Packed in bales of standard dimensions, not less than 54 inches long, approximately 1,000 to 1,500
Moisture:	Packed air dry
Prohibitive Materials:	Less than 1 percent
Outthrows:*	Less than 5 percent
Water Solubles:	Less than 2 percent of the acceptable corrugated (Note a)
Organic Solubles:	Less than 2 percent of the acceptable corrugated (Note b)
Ash:	Less than 1 percent of the acceptable corrugated (Note c)

Note a: Determined by ASTM D-1162 or equivalent
Note b: Determined by ASTM D-1804 or equivalent
Note c: Determined by ASTM D- 586 or equivalent

* Outthrows are contaminants which make the product unsuitable for comsumption at the grade specified. Outthrows usually consist of materials which are compatible with the paper-making process. Examples include cloth bindings, chipboard, string bindings and glassine.

Source: Alter, H. and Reeves, W.R., Specifications for Materials Recovered from Municipal Refuse, The National Center for Resource Recovery, May 1975, NTIS #PB-242 540.

Exhibit B4

CHEMICAL ANALYSES FOR ALUMINUM SCRAP, MAXIMUM PERCENT BY WEIGHT

Element	Reynolds Metals		ALCOA	
	Grade A	Grade B	Grade I	Grade II
Silicon	0.30	0.50	0.30	1.00
Iron	0.60	1.0	0.70	1.00
Copper	0.25	1.0	0.40	2.00
Manganese	1.25	1.25	1.50	1.50
Magnesium	2.0	2.0	2.00	2.00
Chromium	0.1	0.3	0.10	1.30
Nickel	0.05	0.3	N.A.[a]	N.A.
Zinc	0.25	1.0	0.25	2.00
Titanium	0.05	0.05	0.08	0.30
Bismuth	0.02	0.3	N.A.	N.A.
Lead	0.02	0.3	0.04	0.50
Tin	0.02	0.3	0.04	0.30
Others				
Each	0.04	0.05	0.04	0.04
Total	0.12	0.15	0.12	0.12
Aluminum	Remainder	Remainder	Remainder	Remainder

[a]N.A. - not applicable

Source: Reynolds Metals and ALCOA.

Exhibit B5

WASTE GLASS AS A RAW MATERIAL FOR THE MANUFACTURE OF GLASS CONTAINERS[1]

1. Scope

   1.1 This specification covers particulate glass (cullet) material recovered from waste destined for disposal smaller than 6 mm, intended for reuse as a raw material in the manufacture of glass containers.

2. Applicable Documents

   2.1 ASTM Standards:
   C 162 Definitions of Terms Relating to Glass and Glass Products[2]
   C 429 Sieve Analysis of Raw Materials for Glass Manufacture[2]
   C 169 Chemical Analysis of Soda-Lime and Borosilicate Glass[2]
   E 688 Testing Waste Glass as a Raw Material for Manufacture of Glass Containers[3]

---

[1]This specification is under the jurisdiction of ASTM Committee D-38 on Resource Recovery and is the direct responsibility of Subcommittee E 38.05 on Glass.
Current edition approved Nov. 1979. Published January 1980.
[2]Annual Book of ASTM Standards. Part 17
[3]Annual Book of ASTM Standards. Part 41

Appendix B—Recovered Materials Specifications

3. Definitions

3.1 flint glass cullet - a particulate glass material that contains no more than 0.1 weight % Fe_2O_3 or 0.0015 weight % Cr_2O_3 as determined by chemical analysis.

3.2 For definitions of other terms used in this specification, refer to Definitions C 162.

4. Representative Sample

4.1 The following requirements qualify the glass lot to be used for direct use in soda-lime glass container manufacturing. Sample should be prepared and examined in accordance with Methods E 688.

NOTE 1 - A preponderant proportion of glass cullet will be soda-lime bottle glass, the glass cullet having a composition as follows as determined by Method C 169.

Oxide	Composition Weight %
SiO_2	66 to 75
Al_2O_3	1 to 7
CaO + MgO	9 to 13
Na_2O	12 to 16

NOTE 2 - All percents referred to in this specification are weight percents.

5. General Requirements

5.1 The sample shall show no drainage of liquid and be noncaking and freeflowing. A moisture content of less than 0.5 weight % is required to meet the free-flowing characteristics of a cullet that is predominantly of smaller particle size 1.18-mm (No. 16) sieve or smaller.

5.2 Screen Size - No material shall be retained on a 6-mm (¼-in.) screen. Material not exceeding 15 weight % shall pass through a 106-mm (No. 140) screen.

5.3 Organic Materials - The total content of organic materials, as measured in accordance with Section 6 shall not exceed 0.2 weight % of dry sample, except for color-mixed glass where the content of organic material greater than 0.2 weight % must be held within a tolerance of ±0.05 weight %, with a maximum organic limit of 0.4 weight %.

5.4 Magnetic Materials - The total magnetic materials shall not exceed 0.05 weight % of dry sample weight for flint glass and 0.14 weight % for colored glass of dry sample weight in accordance with Section 6.

5.5 Permissible Color Mix for Color Sorted Glass Cullet by Weight:

5.5.1 Amber Glass Cullet:
- 90 to 100% amber
- 0 to 10% flint
- 0 to 10% green
- 0 to 5% other colors

5.5.2 Green Glass Cullet:
- 50 to 100% green
- 0 to 35% amber
- 0 to 15% flint
- 0 to 4% other colors

5.5.3 Flint Glass Cullet:
- 95 to 100% flint
- 0 to 5% amber
- 0 to 1% green
- 0 to 0.5% other colors

5.5.3.1 Percents above 0.1 weight % of Fe_2O_3 or 0.0015 weight % of Cr_2O_3 or both, as determined by chemical analysis shall be considered mixed color glass. These limits are consistent with industry experience on raw material.

5.5.3.2 Flint glass cullet may contain up to 1 weight % emerald green or 10 weight % Georgia green, or a combination within the limits 1% Georgia green = 0.1% emerald green.

5.6 Other Inorganic Material (such as non-magnetic metals or refractories) - As measured, material larger than 850 mm (No. 20) screen size shall not exceed 0.1 % of the dry sample weight. Material smaller than 850-mm screen size shall not exceed 0.5 % of the dry sample weight.

5.6.1 Refractories - Based upon U.S. series screen size and sample weight, the following refractory particle limits shall apply for each screen fraction as stated below.

+20 mesh	1 particle per 18-kg (40-lb) sample
-20, +40 mesh	2 particles per 450-g (1-lb) sample
-40, +60 mesh	20 refractory particles per 450-g (1-lb) sample

5.6.2 Nonmagnetic Metals:

+20 mesh   1 particle per 18-kg (40-lb) sample

Upon failure to meet the previously stated specification limits, retesting is permissible.

6. Sampling and Testing

6.1 Sampling and testing shall be in accordance with Methods E 688.

Index

ALCOA, 37, 114
ASTM, 35, 36, 37
Accelerated depreciation, 252
Accessibility, 204
Air classification, 68, 80-84
Air knife separator, 105
Air pollutants, 204
Air pollution, 210-213
Air preheater, 87
Air quality, 221
Air quality analysis, 231
Air quality regulations, 227
Akron, OH, 126, 136, 156, 236, 237
Albany, NY, 155
Algicides, 209
Aluminum, 28, 41, 106, 114
Aluminum Company of America, 28, 37
Amber glass, 106
American systems, 89, 131
Americology, 135, 136
Ames, IA, 32, 126, 135, 153
Andco-Torrax, 33, 142, 144
Apron conveyors, 117-121
Architect-engineer, 238, 239
Ash disposal, 224-226
Ash disposal landfill, 225
Atlas silo, 126
Auburne, ME, 153
Autogenous combustion, 184
Avoided costs, 246
Azuza, CA, 160

BACT, 231
Back-end recovery, 132

Baghouse, 214, 215-216
Baltimore, MD, 153
Baltimore County, MD, 153
Baseline air quality, 232
Batesville, AK, 151
Belt conveyors, 121-125
Benzo(a)pyrene, 213
Betts Avenue, New York, NY, 156
Black Clawson, 26, 68, 126, 140, 187
Blowdown water, 209
Blytheville, AK, 151
Boiler blowdown, 208
Bottom ash, 224
Bradley East Landfill, CA, 160
Braintree, MA, 133, 154
Brass, 30
Brea, CA, 160
Bridgeport, CT, 140, 151, 236
Building materials, 30
Burley, ID, 153
Burlingame, VT, 158

CEA, 138
Cuba, NY, 155
Calorimeter, 17
Calumet City, IL, 161
Cans, aluminum, 114
Capacity credit, 246
Carbon monoxide, 96, 211, 213, 214
Carson, CA, 160
Cattaraugus County, NY, 155
Central Contra Costa, CA, 184
Central disposal, 86
Centrifugal spray scrubber, 217

267

Char, 91, 224
Chicago Northwest, IL, 133, 153
Chicago Southwest, IL, 126, 135, 153
Chlorinated hydrocarbons, 230
Cinnaminson, NJ, 161
City of Industry, CA, 160
Coal, 135
Codisposal, economics, 192
Codisposal, status, 187
Cogeneration facility definition, 244
Co-incineration, 180
Collection methods, 195
Collection trucks, 195
Collegeville, MN, 154
Columbus, OH, 157
Combustibles in residue, 225
Combustion calculations, 95-99
Combustion equipment associates, 32, 138
Commuter traffic, 206
Competitive bidding, 242
Construction loan guarantees, 252
Construction permits, 233-234
Continuously discernable, 222
Contract, full service, 241
Contract, hybrid, 241-242
Contract, long-term, 243
Contract, modified full service, 241
Contract, negotiated, 242
Contract non-negotiated, 242
Contractural flow control, 236
Control equipment selection, 221-222
Control technology, 214
Controlled air incineration, 92
Conveyor funneling, 117
Cooling tower blowdown, 208
Cooling water, 208-210
Copper, 30
Copper precipitation, 27
Corona, CA, 160
Costs, collection systems, 197-201
Costs, disposal, 198
Costs, installed capacity, 162
Costs, operation & maintenance, 162
Costs, resource recovery facilities, 162
Crane and bucket, 126
Criteria pollutants, 212, 229
Crossville, TN, 157
Cullet, 29, 42, 101
Cyclone, 115

DDT, 230
DOE, 252
Dade County, FL, 28, 152
De-mineralizers, 209
Denmark, 1
Densified RDF, 136

Denver, CO, 161
Department of Energy, 252
Detinning, 27
Detroit, MI, 154
Diffusion model, 232
Dilution, 210
Dings, Inc., 101
Dioxins, 213, 230
Duarte, CA, 160
Duluth, MN, 155, 187
Durham, NH, 155
Dust RDF, 138
Dyersburg, TN, 157

EIS, 226
EPA emission factors, 211, 212, 213
EPA emissions regulations, 227
EPA funding, 192-193
ESP, 208
East Bridgewater, MA, 154
East Hamilton, Ontario, 126
Eco-Fuel II, 138-140, 208
Eddy current separation, 115-116
Electric power sale, 243-246
Electrical generation, 34
Electrostatic precipitator, 215, 217-220
Electrostatic separation, 117
Elutriation, 114
Energy Security Act, 252
Energy tax credits, 252
Energy users, 48
England, 1
Environmental impact, 204
Environmental impact statement, 226
Eriez Magnetics, 105
European refuse, 1
European systems, 86, 131

FERC, 35, 244
Fabric filters, 208, 214, 215-216
Facility siting, 204
Fenwal, 79
Ferrous metals, 27, 39, 101, 195
Fiberclaim, 26
Fine-tuning, 237
Flail mill, 101
Flash drying, 181
Flash pyrolysis, 140
Flint, 29
Florida, State of, 246
Flow control, 235-236, 242
Flow control ordinances, 236-237
Fluff RDF, 135
Fluidized bed incineration, 185-187
Fly ash, 224
Fly ash removal, 213
Formal bids, 238
Franklin, OH, 187

Frequency, 18
Fresh Kills, NY, 161
Frit, 144
Front-end recovery, 132
Froth flotation, 106
Ft. Dix, NJ, 155
Ft. Eustis, VA, 158
Ft. Knox, KY, 153
Ft. Leonard Wood, MN, 155
Full service contract, 241

GO bonds, 249
Gallatin, TN, 158
Gasifier, 142
Gatesville, TX, 158
General obligation financing, 248–251
Genesee Township, MI, 154
Germany, 1
Glass, 29, 42
Glass recovery, 106–113
Glassphalt, 224
Glen Cove, NY, 156, 181
Glenwillow landfill, 237
Gold in garbage, 3
Grate, 86, 87, 89, 91
Green glass, 106
Groveton, NH, 155

HMS, 40
Hamburg, Germany, 1
Hamilton, Ontario, 136, 159
Hammermills, 69, 71, 73, 101
Hampton, VA, 133, 158
Handpicking, 100
Harrisburg, PA, 133, 157
Harrisonburg, VA, 158
Haverhill, MA, 154
Hazardous air pollutants, 213–214
Heavy media separation, 115
Heavy metals, 208, 230
Hempstead, NY, 26, 28, 126, 140, 156
Herreshoff furnace, 181
Highway, 7
Honolulu, HA, 152
Hooker Energy Corp. 126, 136, 156
Hot water generation, 34
Huntsville, AL, 151
Hybrid contracting, 241–242
Hydrocarbon emissions, 211, 213, 214, 230
Hydrogen chloride, 96, 213
Hydropulper, 140

ISIS, 36
Implementation, conventional, 240–241
Incentive programs, 196
Incineration, 85, 180

Indiana General, 105
Industrial development bonds, 252
Industrial waste, composition, 18
Industrial waste, definition, 11
Industrial waste, generation, 11
Industrial waste, projections, 13
Industrial waste, volume, 12
Industrial waste, weight, 12
Intake structures, 210
Interconnection costs, 246
Interest cost deduction, 252
Internal revenue bonds, 248
Investment tax credits, 252

Jacksonville, FL, 152
Jigging, 114

Kaiser Aluminum, 28

LAER, 231
LIM, 115–116
LOI, 46
Lakeland, FL, 152
Lakes, 209
Land application, 180
Landfill, 4, 180, 225
Landfill, ash disposal, 225
Landfill gas generation, 33
Landguard, 140
Lane County, OR, 157
Lawrence, MA, 154
Lead, 30
Legal constraints, 242–243
Letter of intent, 46
Letter of interest, 44
Leveraged leasing, 248
Lewisburg, TN, 158
Linear induction motors, 115–116
Live bottom, 59
Live bottom bin, 126
Livingston, MT, 155
Los Angeles, CA, 160
Love Canal, 4

MCU, 92
MSW, 211, 213, 222, 224
MSW, definition, 7
MSW, estimates, 8
MSW, projections, 10
MSW, volume, 9
MSW, weight, 8
Madison, WI, 159
Magnetic belt separator, 101
Magnetic drum, 101
Magnetic separation efficiency, 105
Magnetic separators, 101–105
Markets, 44
Markets, energy, 31

Markets, materials, 25
Martinez, CA, 160
Mass burning, 132
Materials purchasers, 47
Median, 18
Methane gas generation, 33
Miami, OK, 157
Micron, 211, 215, 221
Milwaukee, WI, 126, 135, 159
Modelling, 232
Modified full service contract, 241
Modular combustion, 145-150
Monroe County, NY, 28, 156, 236
Monsanto, 33, 140
Monterey Park, CA, 160
Mountain View, CA, 161
Multiple hearth incinerator, 181-185
Municipal revenue bonds, 248

NAAQS, 229-232
NEPA, 226
NPDES, 232-233
NSPS, 227
Nashville, TN, 133, 158
National ambient air quality standards, 229-232
National Environmental Policy Act, 226
Natural gas, 145
Negotiated contract, 242
Negotiated procurement, 238, 239
Netherlands, 1
New Jersey, State of, 236
New Orleans, LA, 153
New source performance standards, 227, 232
New York, 1
Newport News, VA, 158
Newsprint, 41
Niagara Falls, NY, 136, 156
Nitrogen oxides, 96, 211, 213, 214
Noise, 223-224
Non-criteria pollutants, 230
Non-ferrous metals, 114, 195
Non-negotiated procurement, 239
Norfolk, VA, 133, 158
North Andover, MA, 154
North Little Rock, AK, 145, 151

ONF, 41
Occidental Petroleum, 33, 140
Ocean dumping, 180
Oceanside, NY, 133, 156
Odor, 222-223
Olinda landfill, CA, 160
Operating permits, 233-234
Optical sorting, 106
Orange County, FL, 152

Osceola, AK, 151
Oxygen-fed pyrolysis, 142
Oyster Bay, NY, 156

PCB's, 213, 230, 248
PSD, 222, 227, 230-232
PSI, 36
PURPA, 35, 243-246, 252
PURPA/cogeneration, 244
Packer truck, 195, 196
Palestine, TX, 158
Palos Verdes, CA, 161
Paper, 26, 41, 100, 195
Parascrew system, 126
Parkdale, Prince Edward Island, 160
Particulate matter, 211-212, 232
Peekskill, NY, 156
Pellet mills, 136
Petersburg, VA, 159
Pinellas, FL, 152
Pit-and-crane, 57
Pittsfield, MA, 154
Pneumatic conveyors, 125
Pollutants, air, 204
Pollutants, water, 204
Pollution control revenue bonds, 248
Polyvinyl chloride, 213
Pompano Beach, FL, 152
Portsmouth, NH, 133, 155
Portsmouth, VA, 159
Primary standards, 222
Process wastewater, 208
Procurement alternatives, 238
Procurement, non-negotiated, 239
Product specifications, aluminum, 36
Product specifications, ferrous metals, 36
Product specifications, glass, 37
Product specifications, methane, 38
Product specifications, nonferrous metals, 37
Product specifications, paper, 36
Product specifications, RDF, 37
Product specifications, steam, 38
Project financing, 247-248
Public hearing, 234
Public ownership, 241
Public review, 226
Purox, 142-145
Pyrolysis, 32, 33, 91, 140-145

Quads, 1
Qualifying facilities, 244
Queuing, 53

RCRA, 4
RDF, 32, 133-140, 180, 224
RDF storage, 126-129
RFP, 239

Index 271

Rack system, 196
Receiving, 53
Recovery efficiency, 162
Redwing, MN, 155
Refractory-lined, 132
Remelt, 27
Rental cost deduction, 252
Request for proposal, 239
Residential development, 204
Residue disposal, 224-226
Residue quench, 208
Residue volume, 225
Resource recovery facility aesthetics, 222-226
Revenue bonds, 252
Revenue, energy, 42
Revenues, estimating, 39
Reynolds Metal Company, 28, 37
Rivers, 209
Riverview, MI, 161
Road building, 30
Rochester, MA, 154
Rod mills, 101
Rotary kiln, 95

SIC, 11
SWARU, 136
Salem, VA, 159
San Fernando, CA, 161
San Leandro, CA, 161
Saugus, MA, 133, 154, 236
Scales, 51
Scavenger magnet, 101
School, 7
Scrap dealers, 27, 29, 31
Screening, 100
Scrubbers, 216-217
Scrubbing liquids, 216
Secondary roadways, 206
Secondary standards, 222
Separation, 80
Sewage sludge, 180-187
Shredder, explosions, 79
Shredder, particle size, 75
Shredder, power requirements, 75
Shredding, 68-82, 101, 123
Siting, 204, 222-223
Slag, 224
Sludge disposal, 180
Sludge marketing, 180
Smith, Frank A., 8
Solid fuel, 32
Solid waste, composition, 14
Solid waste, estimates, 6
Solid waste, European, 1
Solid waste, heating value, 16
Solid waste, moisture content, 15
Solid waste, sampling, 17
Solid waste, seasonal variation, 14

Solid waste, shortfall, 6
Solid waste, testing, 17
Source impact modelling, 231
Source separation, 29
South Charleston, WV, 145
Southeastern Tidewater Energy Project, VA, 159
St. Louis, MO, 32, 135
Stainless steel, 30
Standard deviation, 18
Starved air systems, 92, 145-150, 184
Staten Island, NY, 161
Statistical analysis, 18
Steam generation, 34
Stearns magnetics, 101
Stoiciometric air, 92
Stoker-fired, 132
Storage, 53
Storage systems, 126-129
Sulfur, 135
Sulfur oxides, 211, 212, 214, 232
Suspension-firing, 91
Sweden, 1
Switzerland, 1
Synfuels bill, 252

Tacoma, WA, 159
Tax savings, 252-253
Thermal discharge, 232
Tipping fee, 236-237
Tipping floor, 58, 59
Toronto, Ontario, 160
Torrax, 33
Traffic, 205-207
Transport vehicles, 224
Transportation peaks, 206
Travelling screens, 210
Travelling-wave separators, 115
Trommel screens, 63, 100
Trough angle, 123
Trucks, 9
Tulsa, OK, 157
Turnkey, 241
Two-stage combustion, 149-150

Union Carbide, 33, 140, 142-145
Union Electric, 135
United States, 1, 150

Variation, coefficient of, 18, 22
Venturi scrubber, 217
Vibrating screens, 63, 101
Visual impact, 224

Walt Disney World, FL, 142, 152
Warwick, RI, 157
Wateshed, 205
Water discharge, 208
Water elutriation, 114

Water pollution, 204, 207-208
Water quality regulations, 232
Waterwall, 221
Waterwall furnace, 132, 133, 208, 213
Waukesha, WI, 159
Weighing facilities, 50
Westchester County, NY, 156
Wet-pulped RDF, 140

Wet scrubber, 214
Williamsburg Bridge, NY, 1
Wilmington, CA, 161
Wilmington, DE, 152
Windham, CT, 152
Winston-Salem, NC, 162

Zinc, 30